Power Electronics and Power Systems

Series editors

The Power Electronics and Power Systems Series encompasses power electronics, electric power restructuring, and holistic coverage of power systems. The Series comprises advanced textbooks, state-of-the-art titles, research monographs, professional books, and reference works related to the areas of electric power transmission and distribution, energy markets and regulation, electronic devices, electric machines and drives, computational techniques, and power converters and inverters. The Series features leading international scholars and researchers within authored books and edited compilations. All titles are peer reviewed prior to publication to ensure the highest quality content. To inquire about contributing to the Power Electronics and Power Systems Series, please contact Dr. Joe Chow, Administrative Dean of the College of Engineering and Professor of Electrical, Computer and Systems Engineering, Rensselaer Polytechnic Institute, Jonsson Engineering Center, Office 7012, 110 8th Street, Troy, NY USA, 518-276-6374, chowj@rpi.edu.

More information about this series at http://www.springer.com/series/6403

Tatsuya Suzuki • Shinkichi Inagaki
Yoshihiko Susuki • Anh Tuan Tran
Editors

Design and Analysis of Distributed Energy Management Systems

Integration of EMS, EV, and ICT

Springer

Editors
Tatsuya Suzuki
Graduate School of Engineering
Nagoya University
Nagoya, Aichi, Japan

Shinkichi Inagaki
Graduate School of Engineering
Nagoya University
Nagoya, Aichi, Japan

Yoshihiko Susuki
Graduate School of Engineering
Osaka Prefecture University
Sakai, Osaka, Japan

Anh Tuan Tran
Graduate School of Engineering
Nagoya University
Nagoya, Aichi, Japan

ISSN 2196-3185 ISSN 2196-3193 (electronic)
Power Electronics and Power Systems
ISBN 978-3-030-33671-4 ISBN 978-3-030-33672-1 (eBook)
https://doi.org/10.1007/978-3-030-33672-1

This Springer imprint is published by the registered company Springer Nature Switzerland AG.
The registered company address is: Gewerbestrasse 11, 6330 Cham, Switzerland

Preface

Smart grid infrastructure has attracted substantial interest for use as effective electric power management systems that work together with recently developed reciprocal communication networks. One of the most important considerations for smart grid systems is maximizing the use of renewable power sources such as wind and solar power, which are beneficial for the economy and sustainability. To achieve a balance of supply and demand with renewable energy sources, a dynamic pricing strategy intended to bring consumer demand more in line with supply has been proposed. However, the success of this strategy strongly depends on actual consumer response to time-varying prices. As can be imagined, having consumers be continuously aware of prices is generally inconvenient and impractical. Therefore, it is highly recommended to have a distributed energy management system (EMS) that can automatically regulate supply–demand imbalances with regard to user convenience and economy.

Additionally, demand for electric vehicles (EVs) and plug-in hybrid vehicles (PHVs) is expected to increase, making them important transportation means in a low carbon society. EVs and PHVs carry in-vehicle batteries of 20–30 kWh and 5–10 kWh, respectively. In addition, vehicles used for car sharing, customer transportation, luggage transportation, home delivery, and other such tasks will be electrified. Thus, there is little doubt that the market for EVs and PHVs will continue to develop. Setting the capacity, number, and arrangement of storage batteries and scheduling battery charging and discharging of batteries are important design points for realizing a distributed EMS that meets various requirements. As EVs and PHVs proliferate, in-vehicle batteries will not be used solely for transportation, but can be integrated into the customer's EMS. Effective use of in-vehicle batteries can be a key technology to realize a resilient community, particularly when supply and demand balance is significantly disrupted or when a disaster occurs. In consideration of these factors, in-vehicle batteries will play three roles in next-generation smart communities.

1. Power storage device for transportation
2. Power storage device for improving supply–demand balance in each customer's local EMS
3. Aggregated community virtual power storage used for power network stabilization

It is highly recommended to design charge–discharge scheduling of in-vehicle batteries to balance supply and demand in the local EMS and meet the community's power network stability requirements while considering the transportation demands of each user. Therefore, an energy–mobility integrated system model must be developed and fully exploited to design the future smart community. Based on these perspectives, this book addresses the following topics:

Part I Design and analysis of energy management systems considering consumer demand and use of electric vehicles

- Activity-based modeling for integration of energy systems for house and electric vehicle

This chapter introduces an activity-based approach to model home and transportation energy demand. This method simulates in-home and out-of-home activities and quantifies home and transportation energy demands accordingly. This method enables quantification of energy demand with realistic temporal variations. An important application of this method is analyzing the EMS that integrates the house and EV or PHV, which can be achieved when in-home and out-of-home activities are modeled consistently.

- Probabilistic model and prediction of vehicle daily use

To utilize in-vehicle batteries in an EMS, with consideration of user acceptance, the EMS must know when the vehicle is driven and parked, which can be represented by a departure, and travel time profile. This chapter presents a method to predict the most probable car use profile over 1 day based on a customer's daily car use statistics. The prediction method is formulated as a maximum-likelihood estimation, and the usefulness of the proposed method is evaluated using numerical experiments.

- Design of a home energy management system integrated to a vehicle (V2H + HPWH EMS)

This chapter proposes a home EMS (HEMS) that simultaneously controls an in-vehicle battery charge–discharge process and a heat pump water heater (HPWH) operation plan. The proposed control method iteratively calculates the in-vehicle battery charge–discharge plan and HPWH operation schedule to minimize the electricity bill considering both reverse surplus power as a penalty and vehicle usage as a constraint. The effectiveness of the proposed system is then verified in simulations using real household data.

- Range extension autonomous driving for electric vehicles based on optimization of velocity profile considering traffic signal information

In this chapter, a range extension autonomous driving system that considers traffic signal information is proposed. This proposed system optimizes the autonomous driving velocity profile based on precise vehicle loss models. The authors conduct simulations and experiments to prove the effectiveness of the proposed system in terms of mileage per charge.

Part II Synthesis of distributed energy management systems based on aggregation of local EMSs and vehicles

- Real-time pricing and decentralized optimization strategy for power flow balancing in EV/PHV storage management

This chapter investigates a decentralized energy management strategy for a community composed of households equipped with EV/PHV storage. The proposed real-time pricing strategy does not depend on the number of agents and allows plug-and-play type operations. This will address EV/PHV storage management problems as unpredictable connections or disconnections of vehicles may exist. Effectiveness of the proposed pricing-based decentralized management strategy is then evaluated using numerical experiments.

- A scalable control approach for providing regulation services with grid-integrated electric vehicles

Researchers at the University of Delaware are providing regulation services by controlling bidirectional power transfer between a fleet of EVs and the grid. As EVs become more popular, increasing the size of the EV fleet, large-scale control becomes an important challenge. Power transfer for thousands of EVs may need to be controlled, accounting for driver requirements while providing regulation services. To cope with this challenge, the authors propose a grid-integrated vehicle (GIV) control approach based on bins. Results show that their GIV control method can combine the good scheduling qualities of a centralized GIV approach with a distributed GIV approach's scalability.

- A continuum approach to assessing the impact of spatiotemporal EV charging to distribution grids

The assessment problem on how charging operations of a large population of EVs impact the spatial voltage profile of a power distribution grid is addressed. Unlike the conventional space-discretization approach, an alternative approach for the assessment based on a continuum representation of the distribution voltage profile is introduced. Continuum representation explicitly keeps spatial (or geographical) information on a distribution grid and thus enables us to quantify the spatial impact of EV charging.

Part III Toward dependable distributed energy management system using information and communication technologies (ICT)

- Cyber security for voltage control of distribution systems under data falsification attacks

This chapter provides an overview of recent research on false data injection (FDI) attacks on voltage measurements transmitted by sensors to a centralized controller. The approach is to first introduce an attack detection algorithm and then enhance the security level using a resilient controller that can perform regulation in the presence of attacks by utilizing detection results. We illustrate our methods using simulation studies on a realistic, small-scale distribution system.

- Machine learning based intrusion detection in control system communication

This chapter compares conditional random field-based intrusion detection with probabilistic model-based intrusion detection. These methods use control system communication network traffic sequence characteristics. Learning only utilizes normal network traffic data, assuming no prior knowledge on attacks in the system. We apply these two probabilistic models to intrusion detection in DARPA data and an experimental control system network and compared performance differences.

Lastly, this work was supported by many collaborative researchers and executed as one of the projects in the Core Research for Evolutional Science and Technology (CREST) program of the Japan Science and Technology Agency (JST), and we deeply appreciate these contributions.

Nagoya, Japan — Tatsuya Suzuki
Nagoya, Japan — Shinkichi Inagaki
Sakai, Japan — Yoshihiko Susuki
Nagoya, Japan — Anh Tuan Tran
May 2019

Contents

Part I
Design and Analysis of Energy Management Systems Considering Consumer Demand and Use of Electric Vehicles

Chapter 1
Activity-Based Modeling for Integration of Energy Systems for House and Electric Vehicle

Yohei Yamaguchi, Nikhil Prakash, and Yoshiyuki Simoda

1.1 Introduction

Energy is consumed to sustain everyday human activities. People use appliances for various purposes such as cooking and entertainment, and use cars to travel to work and other places. Thus, home and transportation energy consumption has been understood, analyzed, and modeled in relation with people's activity. Activity-based modeling generally assumes that energy consumption is derived from the activity of simulated individuals. Thus, activity is dealt with as a core of modeling, which enables replication of the structure determining energy demand. According to Sivakumar [58], activity-based modeling is distinguished from agent-based modeling simulating the actions of individuals or a group of individuals because not all agent-based models are activity-based. In this chapter, we present activity-based modeling techniques to model energy systems integrating houses and electric vehicles.

There have been several activity-based models developed to model home and passenger transportation energy demand. Integrated modeling between homes and passenger transportation can be established using consistent activities people engage in inside and outside their homes. Most existing activity-based models have not been applied to integrated analysis between home and transportation because these domains are developed independently. However, there is an increasing need for integrated analysis for various purposes, such as design of home energy management systems (EMSs) for minimizing household electricity cost because of the emergence of electric vehicles (EVs). In the power system management field, EVs are recognized as a source of flexibility to levelize electric power demand and

Y. Yamaguchi (✉) · N. Prakash · Y. Shimoda
Graduate School of Engineering, Osaka University, Osaka, Japan
e-mail: yohei@see.eng.osaka-u.ac.jp

T. Suzuki et al. (eds.), *Design and Analysis of Distributed Energy Management Systems*, Power Electronics and Power Systems,
https://doi.org/10.1007/978-3-030-33672-1_1

improve power system performance [6]. Several studies have analyzed integrated home EMSs considering EVs [14, 22, 42–44, 67].

Existing activity-based models are characterized by methods in which activity and appliance and car use among other factors are modeled. This chapter introduces a method for classifying activity-based energy demand modeling techniques developed for home energy systems (Sect. 1.2) and transportation energy systems (Sect. 1.3). Section 1.4 presents an activity-based model developed for integrated analysis between homes and EVs.

1.2 Energy Demand Modeling for the Residential Sector

Household electrical appliances can be classified into three groups based on their use [17]. The first group operates all day without occupant intervention such as refrigerators and network routers. The second group includes appliances operated by occupants for certain activities, such as washing machines and televisions. The third group involves heating, air conditioning, and lighting, which are operated to control the indoor environment depending on home occupancy. Based on this understanding, appliance energy consumption has been modeled to realize a relationship with occupancy and activity of people in the house. Established models use a variety of methods to generate occupancy and activity and determine appliance use as explained below.

1.2.1 Occupancy-Based Approach

Richardson et al. [54] established an occupancy-based method for modeling residential energy demand, which has been frequently used to model for dynamic energy demand for residential buildings. The number of active occupants is considered as transition states and a random trial is conducted at each time step to determine the occurrences of transition between transition states. For this process, the Markov property is assumed as the transition probability defined as $N_{i,j}/N_i$, where N_i is the total number of samples at transition state i and $N_{i,j}$ is the number of samples whose state changes from i to j. Transition probability is defined for each time step, and by evaluating with random numbers, a time series of changes in the number of active occupants is first determined.

Time use data (TUD) is generally used in this process. TUD is generally collected by time use surveys, in which respondents are asked to submit time use diaries that describe how they spent their time on survey days. In their model, TUD is classified by respondent household size of when transition probability is quantified.

After generating an occupancy schedule, occupant appliance use is determined. For appliances in the aforementioned second category, a so-called switch-on probability, which is the probability of a switch-on event of an appliance occurs, is used. In Richardson et al. [54], the switch-on probability was quantified using appliance TUD and annual total electricity consumption. They assumed the switch-on probability to be proportional to the probability of activities being undertaken (referred to as activity probability) quantified for each time of day based on the TUD corresponding to the number of active occupants. It should be noted that activity probability is different from switch-on probability. If activity probability is used as switch-on probability, appliance use will be overestimated because switch-on events do not occur at every time step when activity is undertaken [69]. A calibration scalar is thus used to avoid such overestimation. Switch-on probability is defined as the product of activity probability and the calibration scalar, which adjusts the total number of switch-on events per year to replicate the appliance's annual total electricity consumption.

In Richardson et al. [53], this approach was applied to model lighting use. The only difference in this approach was that the lighting model considered natural light depending on outdoor irradiance. This approach can be applied to appliances categorized in the third category operated to control the indoor environment depending on occupancy. This approach has several applications.

Richardson's model has been further extended. For instance, McKenna and Thomson [41] integrated modules to quantify energy demands for both water heating and space heating to cover energy demand for all end-uses. Baetens and Saelens [3] proposed accounting for occupancy patterns [2] to classify TUD, which was originally classified solely by household size. This contributed to improve the model's ability to represent heterogeneity among households. Yamaguchi et al. [69] proposed using a calibration scalar dependent on time of day to improve the representation of appliance use time variation.

One of the shortcomings of Richardson's model is that variations in the number of switch-on events per day cannot be replicated because switch-on events occur as a result of random trials made at each time step. Flett and Kelly [18] overcame this weakness by first determining the number of switch-on events for a simulated day based on empirical data. Switch-on events are then allocated to the timeline considering occupancy.

1.2.2 Activity-Based Approach

The second type of model explicitly simulates household member activities, such as watching television and cooking. Simulated activity classification depends on available TUD. Activities are then converted to switch-on event occurrences. There are two approaches that can be considered for this model type. First, Widen and Wackelgard [64] and Widen et al. [65] proposed a discrete-time Markov chain model in which activities are defined as transition states. Random trials of transitions from one state to another are conducted in each time step.

Secondly, Wilke [66] proposed a discrete-event model in which household member activities are simulated by repeating the following two processes: selecting an activity starting from the examined time of day and selecting the selected activity's duration. For activity selection, the starting probability, at which each considered activity starts, is calculated using multinomial logit models developed for each time of day. Markov chain transition probability can also be used to model the sequence of discrete activities as transition probabilities from one activity to another and is quantified based on TUD as a semi-Markov process. Tanimoto and Hagishima [62] proposed to use the activity probability. Yamaguchi and Shimoda [68], whose work is explained in detail in Section 4, proposed a similar approach and compared how model design influences activity modeling performance.

An appliance switch-on event occurrence is examined in relation to the stochastically determined activity, and both discrete-time and discrete-event trials can be adopted for this process. In discrete-time trials, switch-on event occurrence is examined at each time step during an activity. In a discrete-event trial, occurrence is examined only once during an activity, such as at the activity's beginning or end.

The most important difference between occupancy-based and activity-based approaches is that in an occupancy-based approach, appliance use is modeled independently from other appliances, thus possibly resulting in an unrealistic appliance use sequence. One extreme example is that all appliances in a house are operated simultaneously. In contrast, such unrealistic sequences do not occur in activity-based approaches, as a sequence of activities is generated before a switch-on event occurrence is examined.

1.2.3 Time-Based Household Energy Demand Model

There is another model variety that does not simulate occupancy and activity. In these models, time represents household member activity by assigning appliance switch-on probabilities as a time-dependent quantity. To construct a model, switch-on event occurrence is identified from measurements of appliance power demands (i.e., empirical data). For example, Yilmaz et al. [70] constructed a cumulative distribution function (cdf) for the number of switch-on events for a simulated household. The number of switch-on events is first assigned each day using this cdf. Then the switch-on event occurrence time is randomly determined based on the switch-on probability, which is the sum of observed "switches on" divided by the total number of days. This improves modeling accuracy in terms of the daily number of switch-on events and includes variations in the number of switch-on events for different days within the same household. Paatero and Lund [46] and Gruber et al. [23] developed similar models.

1.2.4 Important Factors in Energy Demand Modeling

Yamaguchi et al. [69] conducted a cross-analysis of these existing household appliance use modeling methods and found that model performance depends heavily on modeling contexts. A model developed based on empirical data of appliance use can replicate household-specific characteristics in appliance use, including both intra/inter-household variation as these variations can be captured from empirical data. However, empirical data availability is limited in many modeling contexts, which is critical when the developed model is applied to contexts outside of where empirical data were collected. TUD-based models have advantages in this aspect, as their analyses revealed that demographics and household characteristics significantly influence household appliance use. Thus, considering their significance will improve model performance. The ability of TUD-based models to reflect demographics and household characteristics enables models to be applied to areas where such conditions are available. In contrast, it is difficult to address cross-area variation in models based on empirical data if empirical data lacks information on demographics and household characteristics.

1.3 Energy Demand Modeling for Transportation

Activity in transportation demand modeling is modeled as a sequence of in-home and out-of-home activities [58]. Activity-based transportation implies that travel derives from daily activities. For example, a person leaves home for work in the morning and travels to an office. That person then leaves the office in the evening and travels back home. Typical activity-based models simulate such sequences, including timing, location, and mode of travel. Transportation energy demand can thus be quantified based on this information.

Transportation research has a long history of modeling travel demand considering the influence of socioeconomic conditions and land-use configuration [4]. Earlier travel demand research focused on evaluating long-term investment-based capital improvement to execute regional planning strategies. Since 1970, attention has shifted to short-term behavior analysis to understand travel schedules of individuals to frame congestion management policies.

Travel demand modeling can be broadly classified into two categories: trip-based modeling and activity-based modeling. Trip-based modeling relies on a top-down nature as it uses overall person-trips in the studied area and is disaggregated into those with different trip origins, destinations, modes, routes, and so on. In contrast, activity-based modeling relies on a bottom-up nature as travel demand is modeled as an aggregation of trips made by independently modeled individuals.

1.3.1 Trip-Based Modeling

Trip-based modeling evolved tremendously in the early days of travel demand modeling [13]. This approach uses trip data collected by individuals, known as person-trip data, and consists of four sequential processes: trip generation, trip distribution, modal split, and trip assignment [45]. Trip generation estimates the number of person-trips emanating from or terminating at an aggregated level of zones. The trip distribution process distributes person-trips from each origin to various destinations. The modal split process assigns travel modes to person-trips with origin-destination pairs. Lastly, the trip assignment process assigns network links or routes to each person-trip. The results illustrate traffic volume on all roads in a traffic network link. This four-stage approach has been widely used for congestion management in transportation planning.

1.3.2 Activity-Based Modeling

The travel activities of individuals can be identified in terms of time, origin, destination, purpose, mode, and route of travel. An activity-based modeling approach considers the decision to travel as a choice among alternatives known as a discrete choice set with alternatives that are exhaustive and mutually exclusive. Discrete choice models are used to model such decision-making and predict the likelihood of an alternative to be selected based on certain input variables. Examples of the most frequently used models are multinomial logit, nested, and mixed logit models [1, 7, 16, 28, 34, 35]. A hazard duration-based model is also often used to model the time until an event is of interest. Mannering et al. [37] developed hazard-based models to analyze time spent by individuals at home before conducting another trip. Bhat and Koppelman [5] analyzed the duration of shopping on the way to work.

Activity-based approaches consider travel a derived demand. The most important feature of discrete choice models is that various factors' influence can be reflected in predicted likelihoods. The seminal works of Chaplin [8], Hagerstrand [24], and Cullen and Godson [11] were based on travel activity analysis and revealed that travel demand is complex and elucidates the relationship between land-use and transportation demand in association with constraints individuals encounter during a particular travel activity. The following factors have been recognized as significantly influencing many papers: socio-demographics and household type and [31, 36, 38–40, 47, 61, 63], land-use and built environment and [12, 16, 32, 49, 55], and travel conditions, including activity duration and [19, 33, 50].

Structural equation modeling, a multivariate statistical technique, is also used for activity modeling. This technique assumes a hierarchical structure among influencing factors, which are classified as exogenous and endogenous variables. This approach considers the influence of exogenous variables on endogenous variables. Activity-based travel demand models are used extensively as this approach captures

(1) direct relationships between activity demand and the need to travel and (2) relationships between participations in various activities. It is also advantageous in modeling joint demand for activity duration and travel [21, 29]. Lu and Pas [36] developed a structure equation-based activity-based travel demand model to analyze the influences of socioeconomic variables on in-home and out-of-home activities.

Conducting higher activity participation within a stipulated time period requires the combination of several trips in a single journey. Such a sequence of activities is described by a trip chain [25, 57]. Henser and Reyes [25] considered seven different trip chains that originated and ended at home that had simple to complex combinations of work and non-work travel alternatives to demonstrate the influence of trip chains on the likelihood of using public transit. Pitombo et al. [48] developed 14 distinct trip chain combinations using activity sequence (home, work, school, and other activities), travel mode sequence (private, public transit, and non-motorized), and the distance between the destination and home traffic zone centroids. This approach is beneficial as energy demand can be quantified more accurately if data describing trip chains are available.

In recent years, interpersonal dependencies among household members have been recognized as a significant factor in modeling travel demand as some travel activities are conducted jointly with other members of the family. In their research, Srinivasan and Athuru [59] investigated the activity allocation and participation of household members for maintenance activities, examining whether an activity is conducted alone or jointly with other household members. They found that gender, working status, household role, and presence/absence of children were significant factors determining joint or solo travel. Srinivasan and Bhat [60] examined the role of intra-household interactions on time consumed by household heads on maintenance activities and household chores in nuclear family households. They observed that gender plays a significant role in executing in-home activities, with non-working women undertaking a considerable share of such activities and that resource constraints such automobile availability are of significant importance in deciding whether an out-of-home maintenance activity is undertaken jointly or individually. An interesting outcome of Kato and Matsumoto [27] revealed that having more children in a household increases the likelihood of conducting joint husband and wife out-of-home leisure activity. Additionally, more non-working days for husbands tends to lower the likelihood of individual out-of-home leisure activities.

Some models constituting sets of rules to establish condition-action pairs have been implemented on an urban scale to schedule traffic volume in congestion management-related studies. CARLA [9] and STARCHILD [51] and [52] were among some of the earliest scheduling models developed for this purpose. CARLA uses a complex combinatorial algorithm based on spatio-temporal and interpersonal constraints to model scheduled activities and their associated durations to decide viable activity patterns. STARCHILD also employs a combinatory approach to select activities before applying a logit choice model to devise the highest utility activity pattern choices. Other models that have been widely accepted are SCHEDULER [20], SMASH [15], and AMOS [30].

As described above, the activity-based approach models travel demand as an accumulation of trips made by individuals. The advantages of an activity-based approach can be summarized as follows: (1) the relationship between trips and factors such as demographics and land-use accessibility can be addressed; (2) the approach provides an effective way to analyze travel demand by aggregating demand considering factors such as travel purpose, travel mode, and locations; (3) it enables consideration of complex behavior such as intra/inter-households activity participation and joint travel decision-making; (4) temporal resolution is well-defined, with activities spread over 24 h in a continuous manner; and (5) travel and energy demands are quantified at various scales from a single house to a community to a nation.

1.4 Case Study

This section presents an activity-based modeling example in which household member activities are randomly simulated for integrated analysis considering a house and an EV. In the analysis, household member daily activities, including those undertaken inside the house and outside the house, are simulated, enabling the energy demand of the house and EV to be quantified in a consistent manner. Activities conducted inside the home such as sleeping, laundry, and cooking are designated as "in-home" activities, whereas activities conducted outside of the home such as leisure outings, going to the supermarket, and commuting to the office are designated as "out-of-home" activities. In the model, out-of-home activity is categorized into two sub-types: home-work-home (HWH) and home-other-home (HOH) as listed in Table 1.1. HWH refers to daily travel related to job, work, or school that originates and terminates at home after conducting the activity. HOH signifies activities other than HWH that originate from and terminate at home.

Figure 1.1 provides an overview of the modeling framework, which consists of the household generation model, the in-home activity model, and out-of-home activity model. The household generation model generates simulated households living in a simulated area based on population census data published in e-Stat [56], which has various datasets available at different spatial boundaries from postcode zone, city, and prefecture to nationwide [26]. Generated households are modeled as a combination of household members to which specific demographic conditions, listed in Table 1.2, are assigned. To assign demographic conditions and combinations of household members, various probability distributions were

Table 1.1 Definition of out-of-house activities

Purpose of travel	Description
Home–work–home (HWH)	Going for work/school and returning back to home
Home–others–home (HOH)	Going for other activities and returning back to home

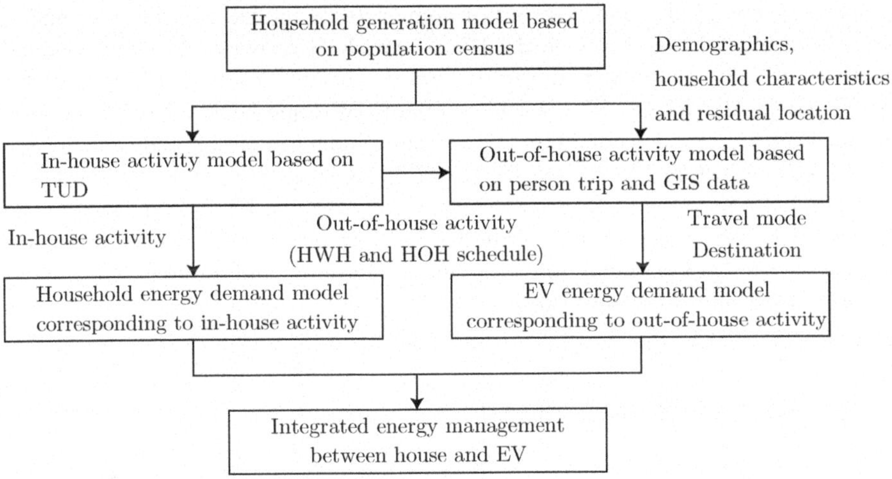

Fig. 1.1 Framework for the activity-based modeling

Table 1.2 Demographics and household characteristics

Category	Item	Note
Demographics	Age	Categorized with 5-year age range
	Gender	Male or female
	Employment type	Full-time, part-time, and unemployed
	School status	Preschool, elementary, junior, and high school, and university
	Occupation	Occupational category
Household characteristic	Member composition	One-person, couple, couple with children, one person with children, couple with parents and children
	Existence of children	Existence of school child, Existence of preschool child
	Household size	Number of household members
House specification	Ownership	Owner or renter
	Housing type	Single detached house or multi-family building
	House size	Total floor area of house
Residence location	City size	Population size of city
	Land use and accessibility to public transport of the area	Population, bus stop and train station density (number of unit per land area)

developed based on population census data. The conditions listed in Table 1.2 are used to select a dataset for the in-home activity model that generates in-home HWH and HOW activities with 5 min intervals for a given period. The in-home activities profile is used in the household energy demand model that quantifies the simulated household's energy demand. The out-of-home activity model uses the information listed in Table 1.2 together with the HWH and HOH activity schedules obtained using the in-home activity model to determine whether or not an EV is used for HWH and HOH activities. The consideration of HWH and HOH activity schedules is necessary because trip conditions significantly influence travel mode choice (see Sect. 1.4.2). When an EV is used for HWH or HOH activities for any household member, the EV is absent from the house. Otherwise, the EV is considered to be connected to the house so that the home EMS or power grid can use the EV's battery. The following subsections describe the in-home and out-of-home activity models and results of the case study.

1.4.1 In-Home Activity Model

Each household member's daily in-home activity is stochastically generated using the model of Yamaguchi and Shimoda [68], which is a discrete-event model in which each household member's activity is separately simulated. In this model, activities are divided into routine and non-routine activities. Routine activities are sleeping, work, or school, commuting to and from school or work, eating meals, and bathing. We refer to these activities as routine because they are assumed to be undertaken once a day routinely in most households. Figure 1.2 illustrates the procedure for generating activities. Before performing the simulation, several modeling parameters such as $PB_{t,n}$ (the percentage of behavior n undertaken at time t) and $DR_{n,du}$ (the cumulative probability of its duration du) are assigned to each household member based on demographic factors such as gender and employment/school status (shown on the left side of Fig. 1.2). In the model, routine

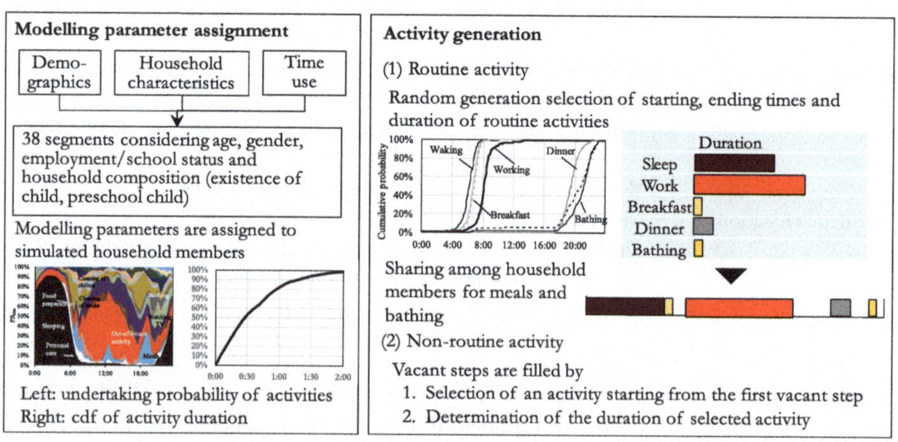

Fig. 1.2 Procedure to generate occupants' presence and activities

activities are first placed on the timeline of a simulated day, and the gaps between routine activities are then filled by non-routine activities.

Routine activities are placed in order of sleeping, work/school, commuting to and from work/school, eating, and bathing. Waking time is first determined by assigning a uniform random number to its cdf. Sleep duration is then determined in the same manner. Using waking time and sleep duration, sleeping can be placed on the timeline. Sleep duration's cdf is conditional on waking time. Other routine activities are placed on the timeline in the same manner considering their dependency on other routine activities.

The distinction between routine and non-routine activities enables the consideration of interactions among household members during routine activities. Interactions among household members occur when they share a space or appliances or coordinate their activity to use them. The model synchronizes household member meal start times. A household member is selected to determine this start time. After determining the mealtime for all household members independently, household member meal start times are shifted to that of the selected person if household members are home at the time. The model also assumed that only one member can use the bathroom at a time. Independent determination of start times may lead to bathing times overlapping among household members. In this case, bathing times of members with start times later than the earliest is shifted. To shift bathing times, the earliest vacant time slot that can contain the determined duration and during which the bathroom is available is found.

It should be noted that aside from sleeping, work/school, and commuting, routine activity occurrence is determined by assigning a uniform random number to routine activity's probability of occurrence (OP_n) because there were TUDs in which routine activities were not undertaken on the day of the survey.

The method used to fill gaps between routine activities is similar to Wilke's model [66]. A non-routine activity is selected for the first vacant slot and its duration is then determined. This process is repeated until all gaps between the routine activities are filled. An activity is selected using non-routine activity ($n = 1$ to N) selection probabilities, $SP_{t,n}$, at time t. The selection probability shows the proportion of activities for which a non-routine activity starting at the examined time slot is selected. The model approximates SP using $PB_{t,n}$ (the percentage of behavior n undertaken at time t) and $DR_{n,du}$ (the cumulative probability of its duration, du). Figure 1.3 illustrates the method for quantifying SP.

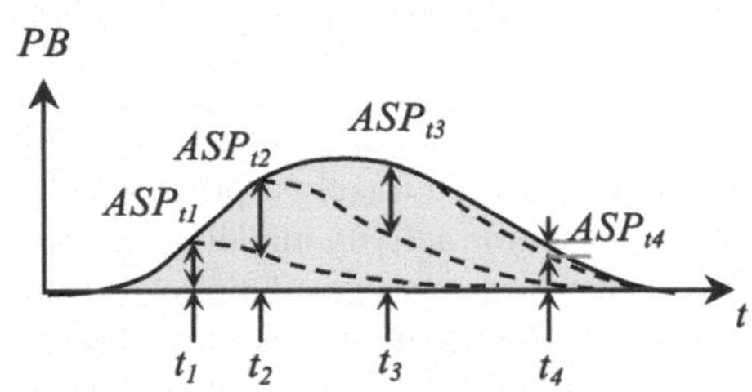

Fig. 1.3 Method to quantify ASP by using PB

Suppose that PB is given for an activity. If the first activity selection is examined at $t = t_1$, ASP_{t_1} is thus given by PB_{t_1}. If this activity is chosen, it has a duration modeled by DR. The portion of PB in the subsequent slots, $t_1 + du$, due to activity n starting at t_1 is given by the product of PB_{t_1} and $(1 - DR_{t-t_1})$, which is illustrated by the first dotted line from the bottom in Fig. 1.3. After an activity is selected based on all activities' SP_t, the next activity selection is examined at $t = t_2$. ASP_{t_2} and is given by the remaining PB_{t_2} after subtracting the dotted area, i.e., $PB_{t_2} - PB_{t_1} \cdot (1 - DR_{t_2-t_1})$. In the proceeding time slots, $ASP_{t,n}$ is given by Eq. (1.1), where K is the number of previous time slots from $t = t_0$ and t_k is the time of the k-th time slot ($k = 1$ to K).

As mentioned above, the model uses several modeling parameters such as PB, DR, and so

$$ASP_{t,n} = PB_{t,n} - \sum_{k=1}^{K} ASP_{t_k,n} \cdot \left(1 - DR_{n,t-t_k}\right). \tag{1.1}$$

$SP_{t,n}$ is given by the proportion of $ASP_{t,n}$ as shown in Eq. (1.2):

$$SP_{t,n} = \frac{ASP_{t,n}}{\sum_{n=1}^{N} ASP_{t,n}} \tag{1.2}$$

These modeling parameters were developed based on the results of the national time use survey conducted by the Statistic Bureau of Japan [56]. That survey used two time use diary formats, Diary A and Diary B, and we used samples of Diary B in this case study. TUD was collected from 18,291 people in 7681 households. Diary B classified time use into 85 activities, which allowed more direct conversion from activities to appliance operation. For example, household maintenance was divided into 14 activities including laundry and cooking activities that could be linked to the operation of washing machines and kitchen stoves, respectively. To reflect the influence of demographics and household characteristics on in-home activity, TUD was classified into 38 categories and modeling parameters were developed for each category, which were assigned to simulated individuals based on the household generation model results listed in Table 1.2.

1.4.2 Out-of-Home Activity Model

As previously mentioned, the model distinguishes out-of-home activity as HWH and HOH based on travel purpose, which was given by the in-home activity model. The out-of-home activity model determines the travel mode and activity destination. This section only presents the model for the travel mode.

We used binomial logit models, shown in Eq. (1.3), to quantify the probability of using the EV, P_{car}, when HWH and HOH are conducted

$$\ln \frac{p_{car}}{1 - P_{car}} = \beta_0 + \sum_{n=1}^{N} \beta_n \cdot x_n, \quad (1.3)$$

where x_n is the n-th predictor variable and β_0 and β_n are the regression coefficients. The predictor variables were those given by the dataset listed in Table 1.3. To develop the logistic regression models for HWH and HOH, the fifth Kinki region person-trip survey (PT) data were used [10]. The survey was conducted to record inter-regional movement of passengers in the Kinki region. The survey period was 2 days, which included 1 weekday and one holiday (Sunday) in the winter. The dataset contains information on five inter-regional modes of travel (air, rail, sea, bus, and car). Additionally, it also contains walking, bicycle, and motorbike travel modes. PT data also includes information pertaining to demographics (age, gender, nature of employment, etc.), household characteristics (household size and composition, head of household, children younger than 5 years old, etc.), and travel conditions (purpose of travel, trip time, distance traveled, arrival and departure time, etc.).

These demographic conditions were used as predictor variables. In addition to demographic conditions, travel condition and residence location were considered in the development of regression models. For travel conditions, departure time and HWH and HOH duration were considered. This information is given by the in-home activity model's results. For residence location and land-use accessibility, three geographical indicators—population, bus stop, and train station densities—were considered. These indicators were quantified for each postcode district in Kinki region.

Continuous predictor variables such as move time, trip time, displacement, out-of-home duration, bus stop density, train station density, and population density are normalized on a logarithmic scale. The underlined predictor variables shown in Table 1.3 are considered as references in the model. Significant predictor variables were determined using the stepwise method that selects predictor variables to minimize the Akaike information criterion (AIC). The results of the regression analysis are as follows:

Table 1.3 Predictors for logistic regression based travel demand model

Category	Predictor variables
Purpose of travel	Gender, age (<30 years, 30–44 years, 45–59 years, 60–74 years, >74 years), person count, person under 5 years old count, nature of employment (self-employed, regular, part-time, executive, contract, temporary), under high school, above high school, possession of car license
Travel conditions	Move time, displacement, trip time, departure time (3:00–6:00, 7:00–10:00, 11:00–14:00, 15:00–18:00, 19:00–22:00, 23:00–2:00), out-of-house duration
Residence location	Number of bus stops per unit area (bus stop density), number of train stations per unit area (train station density), number of people per unit area (population density)

- Gender was insignificant for HWH; however, males have a higher probability of traveling by car than females for HOH travel purposes.
- People older than 45 years old were more likely to travel by car, whereas those younger than 30 were less likely to travel by car for HWH. In the HOH category, all age predictor variables are insignificant except for the 45–59 age group, which is negatively related. This means that people within this age range were less likely to travel by car. Trip time reduces car use probability, whereas total travel time had the opposite effect.
- Occupation was a significant influencer, as self-employed people and company executives have higher car use probabilities for both HWH and HOH travel purposes than temporary, contract, and part-time employees.
- Travel conditions also had significant effects. A person who departs home after 11:00 for HWH activity is less likely to travel by car, whereas those who depart between 3:00 and 6:00 are more likely to travel by car. This may be due to lower frequency of public modes of transportation during early morning hours. In contrast, if a person who departs after 11:00 for HOH activity is more likely to travel by car but is less likely to travel by car for departure times is between 3:00 and 6:00.
- Out-of-home activity duration is indirectly associated with probability of travel by car, meaning that a person who spends more time conducting out-of-home activity (HWH and HOH) is less likely to use the car for travel activities.
- Distance traveled directly influences the probability of travel by car for HWH and HOH activities. In other words, a greater distance covered for out-of-home activities increases car use probability.

The most significant influence was observed for geographical location indicators. Figure 1.4 shows P_{car} values of all postcode districts in the region that were quantified for full-time working male aged 30–44 with the reference category for

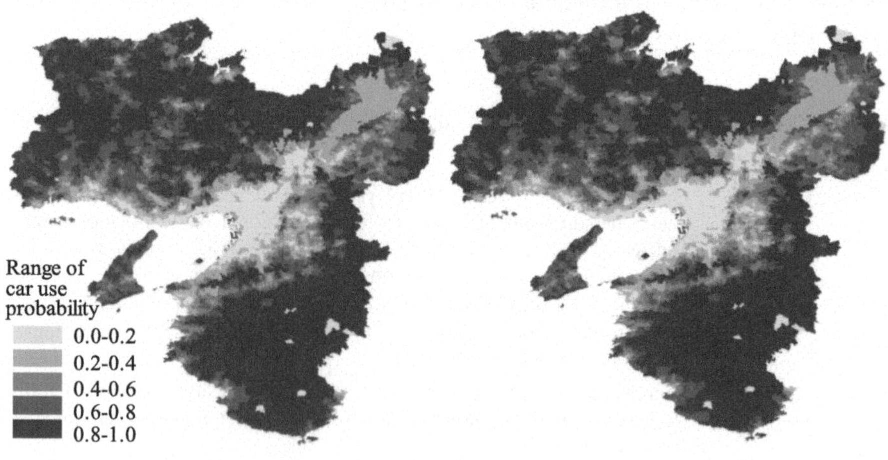

Fig. 1.4 P_{car} in Kinki region: left for HWH; right for HOH

other conditions and a mean value of continuous predictor variables. The map in Fig. 1.4 shows that people living in the central part of the Kinki region have a lower probability of traveling by car for HWH and HOH purposes than those living in the r non-central regions. This is possibly because the central Kinki region has densely populated cities (e.g., Osaka, Kobe, and Kyoto) which have well-developed public transportation infrastructure. Better public transportation connectivity in the central region allows residents to trade-off travel modes between public and private mode of transportation, whereas people living in non-central regions are compelled to drive cars to meet their travel needs because of poor land-use accessibility.

1.4.3 Results

To demonstrate the integration results, activity simulations were conducted for 1 year assuming a four-member household consisting of the people listed in Table 1.4. The floor plan of the residence is shown in Fig. 1.5.

1.4.3.1 Activity

Figure 1.6 shows the composition of in-home, HOH, and HWH for the household members at each time of day. The results were quantified by aggregating simulation results of all days in a year. The results thus contain results of weekdays and holidays. The black areas represent HWH activity, whereas the gray area indicates HOH activity. The remaining portion of Fig. 1.6 shows the proportion of in-home activities. As shown in the figure, all household members have different proportions in the three activities. These differences can be attributed to demographic factors.

Table 1.4 Case study household

Item	Conditions	
Household composition	Working male	Male with a full-time job aged 30–44 years old
	Working female	Female with a part-time job aged 30–44 years
	Primary school student	Male primary school student
	Secondary school student	Female secondary school student
Housing	Residence in a multi-family building with a total floor area of 89 m^2. U value is 0.518 W/m^2. The layout is shown in Fig. 1.5. There are a living room and three bedrooms with an air-conditioner for cooling and heating. Gas condensing boiler is used for water heating	
Residence location	Located in suburban area in Osaka prefecture. The population density is around 10,000 people/km^2, train and bus station densities are 0 stations/km^2 and 0.5 stops/km^2, respectively	

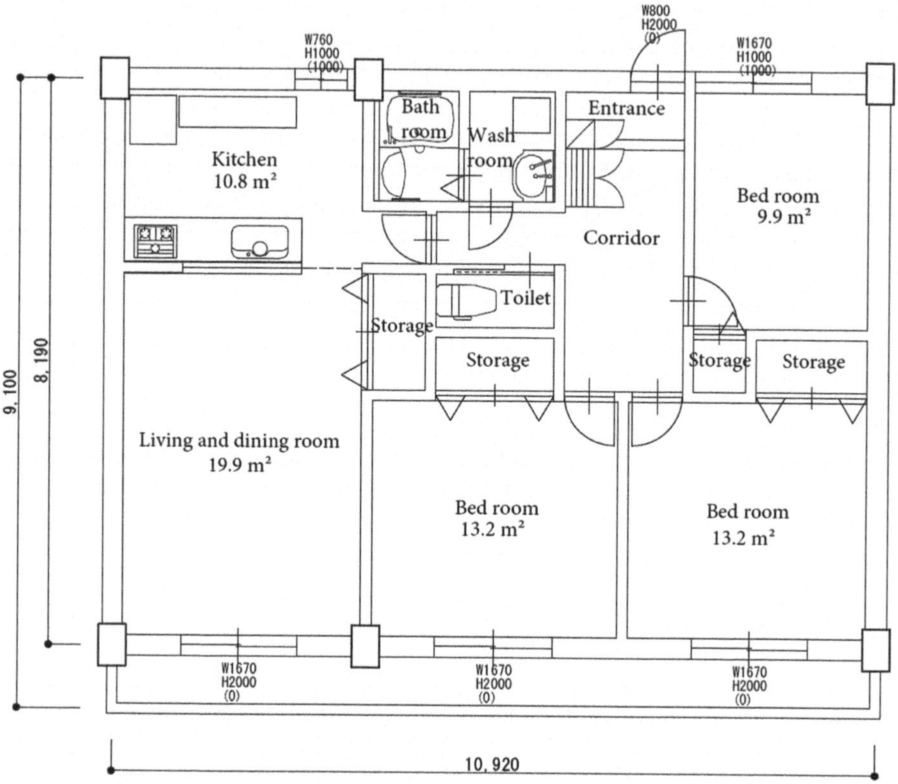

Fig. 1.5 Floor plan of the residence

For example, the working male has a higher HWH probability than the working female, which is mainly due to differences in paid work participation between full-time and part-time jobs.

1.4.3.2 Electricity Demand

Figure 1.7 shows household electricity demand during a week from Monday to Sunday in July. Daytime electricity demand is lower on weekdays than weekends as household members have HWH for work or school on weekdays. The stochastic behavior of electricity demand is generated due to in-home activity. Electricity demand for cooling and lighting varies with meteorological conditions, whereas operating conditions are dependent on occupancy of the conditioned room.

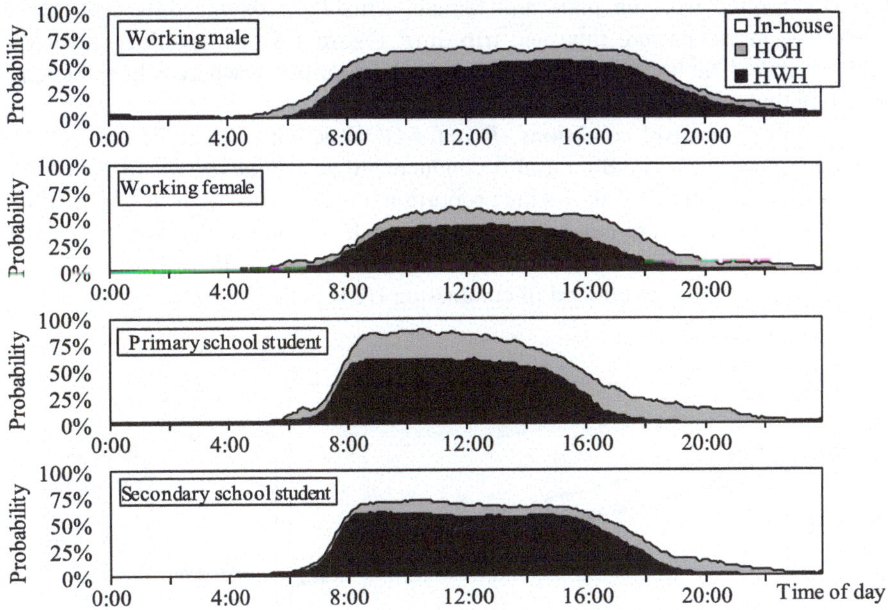

Fig. 1.6 Proportion of in-house, HOH, HWH activities of the household members

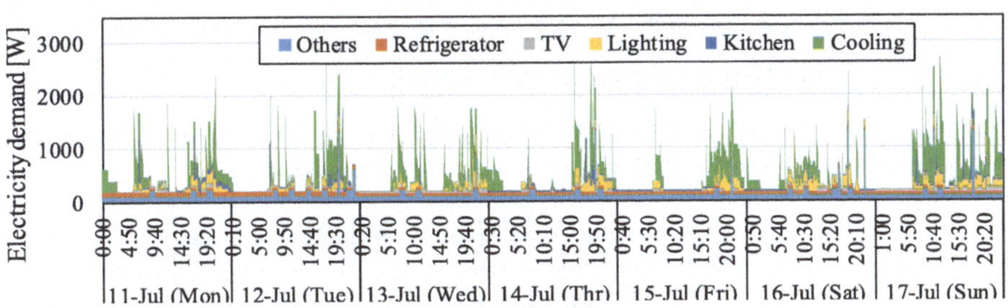

Fig. 1.7 Household electricity demand during a week in July

1.4.3.3 EV Integration Potential

This section discusses EV integration potential, which refers to the likelihood of a car being parked at house and thus available for power interchange between the house and the power grid. Identifying the time of day when an EV can be integrated with a home facilitates the foundation for an EMS between homes and EVs.

In this case study, we assumed that the working male commutes to work by train and that the working female uses the car for HWH, although P_{car} of HWH was quantified at 36% and 32%, respectively. If this model is extended to cover many households, the commuting mode can be randomly selected using a P_{car} estimate for simulated individuals. Random travel mode selection is conducted for all HOH activities for the working male and female, with P_{car} varying only with travel conditions, i.e., departure time and trip time. Figure 1.8 shows influence for the working male. The horizontal axis indicates HOH duration, whereas lines indicate departure times as shown by the graph legend.

Depending on travel conditions, P_{car} for HOH activity can be determined for each outing and then a random trial is conducted to determine the use of the car for these activities. Figure 1.9 shows the proportion of days over 1 year in which HWH and HOH activities were conducted using a car. It was found that working males are more likely to conduct HOH activity in the morning hours than female, whereas females have a higher likelihood of conducting HOH activities in the afternoon.

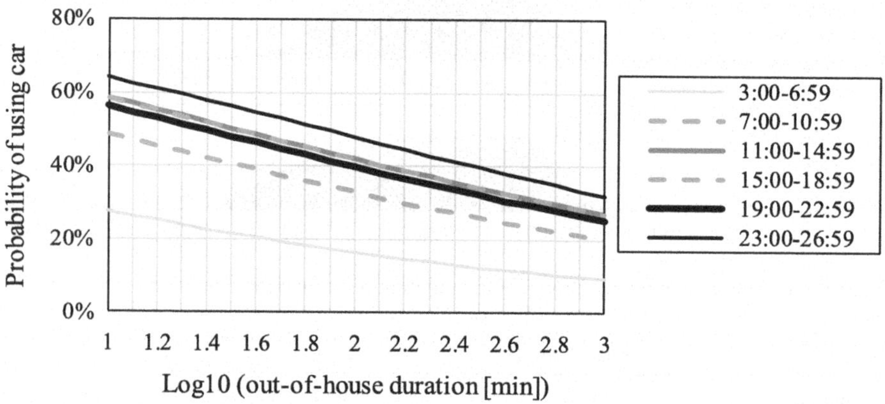

Fig. 1.8 P_{car} of the working male for HOH activities with different travel conditions indicated by the horizontal axis (duration of HOH) and graph legend (departure time)

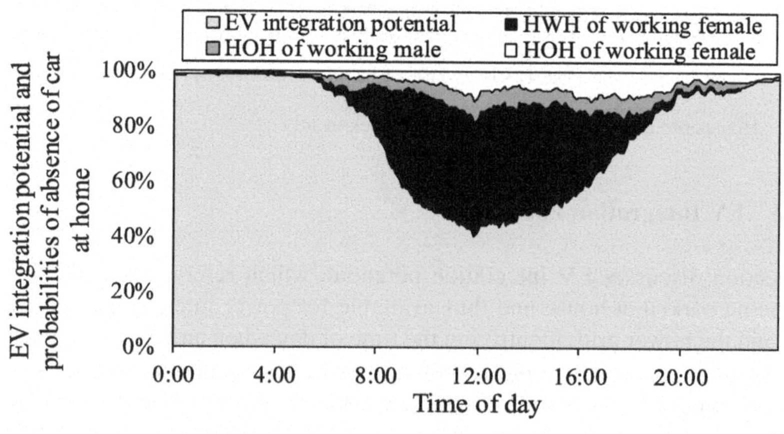

Fig. 1.9 Estimated EV integration potential (bottom part) and probabilities of car use at each time of day for HWH and HOH activities (the remaining parts)

The area below curves in the Fig. 1.9 represents the state of the EV being parked at the house during the day (i.e., EV integration potential). EV integration potential is almost 100% during the early morning hours; however, as the day progresses, the probability of traveling by car increases and EV integration potential subsequently decreases. During the evening hours after work, EV integration potential began to increase.

1.4.3.4 Influence of Geographical Location

It has been discussed in the above sections that geographical location influences people's travel mode choice. To elucidate that, EV integration potential of an urban, suburban, and rural region for the HOH activity in presented in this subsection. The land-use accessibility indicator for urban, suburban, and rural region is mentioned below in Table 1.5.

Figure 1.10 shows EV integration potential of the family with different residence locations listed in Table 1.5. As shown in the figure, EV integration potential is higher in densely populated area due to better access to public transit. The total amount of EV integration potential, during which EV can be used for energy

Table 1.5 Land-use accessibility indicators of urban and rural region

	Location		
	Rural	Suburban	Urban
Population density (people/km^2)	<2000	<14,000	<24,000
Bus stop density (stops/km^2)	0.5	0.75	4
Train station density (stops/km^2)	0	0.5	0.5

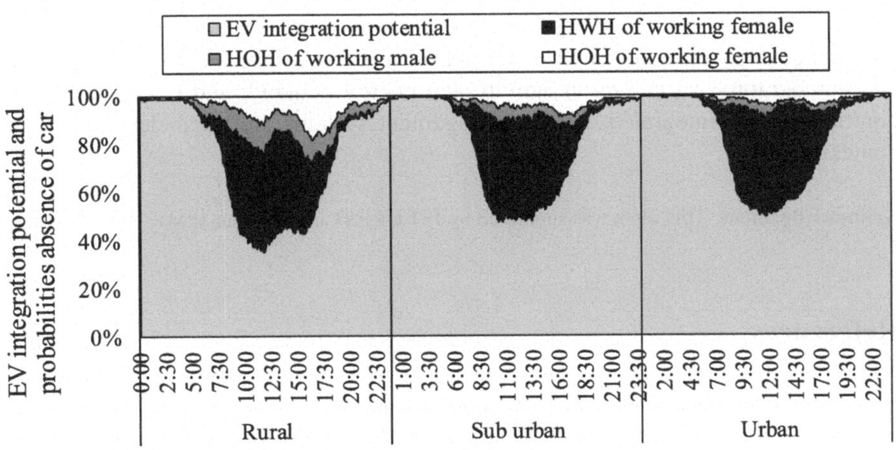

Fig. 1.10 Influence of geographical location on EV integration potential and probabilities of absence of car at home

management, was quantified as 17.6 h/day in the rural area, whereas 19.0 and 19.5 h/day in the suburban and urban areas. This result implies the existence of a large potential for energy management. However, EV integration potential during 9:00–17:00 was quantified as 3.4, 4.2, and 4.5 h in the rural, suburban, and urban areas, respectively. During the time slot, electricity demand is relatively small as shown in Fig. 1.7 and generation of PV is large if it is installed, which indicates that there might be a large scale of mismatch between electricity demand, PV generation and flexibility provided by EV. The mismatch becomes wider if we distinguish weekdays and holidays. Hence, it is also important that EV integration potential and the abovementioned mismatch distribute geographically. Further research is needed to understand the relationship.

1.5 Conclusion

Energy is consumed to sustain people's everyday life. Activity-based energy demand modeling is a computational method to replicate the structure in which energy demand is determined. Various survey data related to people's activity, such as TUD and PT data, have been used to construct models. In the modeling of dynamic behavior of residential energy demand, occupancy or activity has been used to model household level energy demand. In the transportation, travel activity of individuals has been identified in terms of time, origin, destination, purpose, mode, and route of travel, which allows quantifying energy demand for travel activity. Energy management system integrating house and EV can be modeled by using consistent data between in-house and out-of-house activities. This chapter demonstrated such modeling approach by taking a Japanese dataset as a case study. The following areas of study would be relevant for the design of energy management systems integrating houses, EVs, and power systems: (1) the relationship between in-house and out-of-house activities, (2) the geographical and demographic variations in activity and energy demand, and (3) the potential contributions of integrated energy management systems and their dependency in contexts.

Acknowledgement This work was supported by JST CREST under Grant JPMJCR15K4, Japan.

References

1. J.E. Abraham, J.D. Hunt, Specification and estimation of nested logit model of home, workplaces, and commuter mode choices by multiple-worker households. Transportation Research Record, 1606, Paper No. 970686, 2014
2. D. Aerts, J. Minnen, I. Glorieux, I. Wouters, F. Descamps, A method for the identification and modelling of realistic domestic occupancy sequences for building energy demand simulations and peer comparison. Build. Environ. **75**, 67–78 (2014)

3. R. Baetens, D. Saelens, Modelling uncertainty in district energy simulations by stochastic residential occupant behaviour. J. Build. Perform. Simul. **9**(4), 431–447 (2016). https://doi.org/10.1080/19401493.2015.1070203
4. C.R. Bhat, A hazard-based duration model of shopping activity with nonparametric baseline specification and non-parametric control for unobserved heterogeneity. Transp. Res. B **30**, 189–208 (1996)
5. C.R. Bhat, F.S. Koppelman, Activity-based modeling of travel demand, in *Handbook of Transportation Science*, ed. by R.W. Hall. International Series in Operations Research & Management Science, vol. 23 (Springer, Boston, 1999)
6. T. Boßmann, I. Staffell, The shape of future electricity demand: exploring load curves in 2050s Germany and Britain. Energy, 90, 1317–1333 (2015). https://doi.org/10.1016/j.energy.2015.06.082
7. J.L. Bowman, M.E. Ben-Akiva, Activity-based disaggregated travel demand model system with activity schedules. Transp. Res. A **35**, 1–28 (2000)
8. F.S. Chaplin, Free-time activities and the quality of urban life. J. Am. Inst. Planners **37**(6), 411–417 (1971)
9. M.I. Clarke, Activity modelling—a research tool or practical planning technique? in *Behavioural Research for Transport Policy* (VNU Science Press, Utrecht, 1986), pp. 3–15
10. Council for Keihanshin Metropolitan Area Transportation Planning, Fifth person trip survey in Kinki region, 2010. http://www.kkr.mlit.go.jp/plan/pt/research_pt/h22/index.html
11. I. Cullen, V. Godson, Urban networks: the structure of activity patterns. Prog. Plan. **4**, 1–96 (1975)
12. G. Currie, Quantifying spatial gaps in public transport supply based on social needs. J. Transp. Geogr. **18**, 31–41 (2010)
13. J. de Dios Ortuzar, L.G. Willumsum, *Modelling Transport* (Wiley, London, 1990)
14. P. Denholm, M. Kuss, R.M. Margolis, Co-benefits of large scale plug-in hybrid electric vehicle and solar PV deployment. J. Power Sources **236**, 350–356 (2013)
15. D.F. Ettama, A.W.J. Borgers, H.J.P. Timmermans, Simulation model of activity scheduling behaviour. Transp. Res. Rec. **1413**, 1–11 (1993)
16. D. Ettema, T. Schewan, H. Timmermans, The effect of location, mobility and socio-demographic factors on task and time allocation of households. Transportation **34**, 89–105 (2007)
17. S.K. Firth, K.J. Lomas, A.J. Wright, R. Wall, Identifying trends in the use of domestic appliances from household electricity consumption measurements. Energ. Buildings **40**(5), 926–936 (2008)
18. G. Flett, N. Kelly, A disaggregated, probabilistic, high resolution method for assessment of domestic occupancy and electrical demand. Energy Buildings **140**, 171–187 (2017). https://doi.org/10.1016/j.enbuild.2017.01.069
19. X. Fu, Z. Juan, An integrated framework to jointly model decisions of activity time allocation and work-related travel. Transp. Plan. Technol. **40**(6), 689–705 (2017)
20. T. Garling, P. Kwan, G. Golledge, Computational-process modelling of household travel decisions: conceptual analysis and review. Transp. Res. B **28**, 355–364 (1994)
21. T.F. Golob, R. Kitamura, C. Lula, Modeling the effects of commuting time on activity duration and non-work travel, in *Presented at the Annual Meeting of Transportation Research Board* (Washington, 1994)
22. P. Grahn, J. Munkhammar, J. Widen, K. Alvehag, L. Soder, PHEV Home-charging model based on residential activity patterns. IEEE Trans. Power Syst. **28**(3), 2507–2515 (2013)
23. J.K. Gruber, S. Jahromizadeh, M. Prodanovic, V. Rakocevic, Application-oriented modelling of domestic energy demand. Int. J. Electr. Power Energy Syst. **61**, 656–664 (2014)
24. T. Hagerstrand, What about people in regional space? Papers Proc. Reg. Sci. Assoc. **24**, 7–24 (1970)
25. D.A. Henser, A.J. Reyes, Trip chaining as a barrier to the propensity to use public transport. Transportation **27**(4), 341–361 (2000)

26. Y. Kambayashi, Y. Yamaguchi, A.T. Matsuoka, Y. Shimoda, Census and time-use data based community-scale modelling of residential energy demand, in *Proceedings of the 4th Asia Conference of International Building Performance Simulation Association* (2018)
27. H. Kato, M. Matsumoto, Intra-household interaction in nuclear family: a utility maximizing approach. Transp. Res. Board **43**, 191–203 (2009)
28. R. Kitamura, Y.O. Susilo, Is travel demand insatiable? A study of changes in structural relationships underlying travel. Transportmetrica **1**, 23–45 (2005)
29. R. Kitamura, J.P. Robinson, T.F. Golob, M.A. Bradley, J. Leonard, T. Van der Hoorn, A comparative analysis of time use data in the Netherlands and California, in *Proceedings of the 20th PTRC Summer Annual Meeting: Transportation Planning Methods* (1992), pp. 127–138
30. R. Kitamura, I. Pas, V. Lula, K. Lawton, E. Benson, The sequenced activity mobility simulator (SAMS): an integrated approach to modelling transportation, land use and air quality. Transportation **23**, 33–54 (1996)
31. Y. Lee, M. Hickman, S. Washington, Household type and structure, time-use pattern, and trip-chaining behaviour. Transp. Res. **41**, 1004–1020 (2007)
32. J. Lee, J. Nam, S. Lee, Built environment impacts in individual mode choice: an empirical study of the Houston–Galveston metropolitan area. Int. J. Sustain. Transp. **8**, 447–470 (2014)
33. J. Li, J. Weng, C. Shao, H. Guo, Cluster-based logistic regression model for holiday travel mode choice. Proc. Eng. **137**, 729–737 (2016)
34. N. Limtanakool, M. Dijst, T. Schwanen, The influence of socio-demographic characteristics, land use and travel time consideration on mode choice for medium- and longer-distance trips. J. Transp. Geogr. **14**, 327–341 (2006)
35. M. Liu, B. Mao, F. Gao, J. Guo, L. Gao, Analysis on commuter's activity chain choice behaviour, in *Traffic and Transportation Studies* (2008). https://doi.org/10.1061/40995(322)19
36. X. Lu, E.I. Pas, Socio-demographics, activity participation and travel behaviour. Transp. Res. A **33**, 1–18 (1999)
37. F. Mannering, E. Murukami, S.G. Kim, Temporal stability of travelers' activity choice and home stay duration: some empirical evidences. Transportation, **21**, 371–392 (1994)
38. T.K. Mattsson, Children's (in)dependent mobility and parents' chauffeuring in the town and the countryside. J. Econ. and Soc. Geogr. **93**, 443–453 (2002)
39. N.C. McDonald, Does residential density affect the travel "gender gap", in *Conference Proceedings 35: Research on Women's Issues in Transportation* (2005), pp. 68–75
40. N. McDonald, Exploratory analysis of children's travel patterns. Transp. Res. Rec. **1977**, 1–7 (2006)
41. E. McKenna, M. Thomson, High-resolution stochastic integrated thermal-electrical domestic demand model. Appl. Energy **165**, 445–461 (2016). https://doi.org/10.1016/J.APENERGY.2015.12.089
42. P. Mesaric, S. krajcar, Home demand side management integrated with electric vehicles and renewable energy sources. Energy Buildings, **108**, 1–9 (2015)
43. J. Munkhammar, J. Widen, J. Ryden, On a probability distribution model combining household power consumption, electric vehicle home-charging and photovoltaic power production. Appl. Energy **142**, 135–143 (2015)
44. J. Munkhammar, J.D.K. Bishop, J.J. Sarralde, W. Tian, C. Choudhary, Household electricity use, electric vehicle home-charging and distributed photovoltaic power production in the city of Westminster. Energy Buildings **86**, 439–448 (2015)
45. N. Oppenheim, *Urban Travel Demand Modelling: From Individual Choice to General Equilibrium* (Wiley, London, 1995)
46. J.V. Paatero, P.D. Lund, A model for generating household electricity load profiles. Int. J. Energy Res. **30**(5), 273–290 (2006)
47. E.I. Pas, The effect of selected sociodemographic characteristics on daily travel-activity behaviour. Environ. Plan. A **16**, 571–581 (1984)
48. C. Pitombo, A. Sousa, L. Filipe, Classification and regression tree, principal component analysis and multiple linear regression to summarize data and understand travel behaviour. Transp. Lett. **1**(4), 295–308 (2009)

49. C.S. Pitombo, E. Kawamoto, A.J. Sousa, An exploratory analysis of relationships between socioeconomic, land use, activity participation variables and travel patterns. Transport Policy **18**, 347–357 (2011)
50. C. Raux, T. Ma, I. Joly, V. Kaufmann, E. Cornelis, N. Ovtracht, Travel and activity time allocation: an empirical comparison between eight cities in Europe. Transp. Policy **18**, 401–412 (2011)
51. W. Recker, G. McNally, S. Root, A model of complex travel behaviour: Part I. Theoretical development. Transp. Res. A **20**, 307–318 (1986)
52. W. Recker, G. McNally, S. Root, A model of complex travel behaviour: Part II. An operational model. Transp. Res. A **21**, 319–330 (1986)
53. I. Richardson, M. Thomson, D. Infield, A. Delahunty, Domestic lighting: a high-resolution energy demand model. Energy Buildings, **41**(7), 781–789 (2009)
54. I. Richardson, M. Thomson, D. Infield, C. Clifford, Domestic electricity use: a high-resolution energy demand model. Energy Buildings, **42**(10), 1878–1887 (2010)
55. T. Saghapour, S. Moridpour, R.G. Thompson, Modelling access to public transport in urban areas. J. Transp. Geogr. **54**, 273–285 (2016)
56. SBJ, Statistic Bureau of Japan, Ministry of Internal Affairs and Communications, http://www.stat.go.jp/data/kokusei/2010/. Accessed 20 July 2018
57. Y. Shiftan, Practical approach to model trip chaining. J. Am. Plan. Assoc. **1645**, 17–23 (1998)
58. A. Sivakumar, Activity-based modelling for urban energy systems, in *Urban Energy Systems: An Integrated Approach*, ed. by J. Keirstead, N. Shah (Routledge, Abingdon, 2013), pp. 204–217
59. K.K. Srinivasan, S.R. Athuru, Analysis of within-household effects and between-household differences in maintenance activity allocation. Transportation **32**, 495–521 (2005)
60. S. Srinivasan, C.R. Bhat, Modeling household interactions in daily in-home and out-of-home maintenance activity participation. Transportation **32**, 523–544 (2005)
61. Y. Sun, E. Waygood, K. Fukui, R. Kitamura, Built environment or household life-cycle stages—which explains sustainable travel more? Transp. Res. Rec. **2135**, 123–129 (2009)
62. J. Tanimoto, A. Hagishima, Total utility demand prediction system for dwellings based on stochastic process of actual inhabitants. J. Build. Perform. Simul. **3**(2), 155–167 (2010)
63. E.O.D. Waygood, What is the role of mothers in transit-oriented development? The case of Osaka-Kyoto-Kobe, Japan, in *Transportation Research Board Conference Proceedings*, vol. 2 (2011), pp. 1073–1652
64. J. Widën, E. Wackelgard, A high-resolution stochastic model of domestic activity patterns and electricity demand. Appl. Energy **87**(6), 1880–1892 (2010)
65. J. Widën, A. Molin, K. Ellegard, Models of domestic occupancy, activities and energy use based on time-use data: deterministic and stochastic approaches with application to various building-related simulations. J. Build. Perform. Simul. **5**(1), 27–44 (2012)
66. U. Wilke, Probabilistic bottom-up modelling of occupancy and activities to predict electricity demand in residential buildings. Thesis 5673, 2013. https://infoscience.epfl.ch/record/183853/files/EPFL_TH5673.pdf
67. X. Wu, X. Hu, S. Moura, X. Yin, V. Pickert, Stochastic control of smart home energy management with plug-in electric vehicle battery storage and photovoltaic array. J. Power Sources **333**, 203–212 (2016)
68. Y. Yamaguchi, Y. Shimoda, A stochastic model to predict occupants' activities at home for community-/urban-scale energy demand modelling. J. Build. Perform. Simul. **10**, 565–581 (2017)
69. Y. Yamaguchi, S. Yilmaz, N. Prakash, S.K. Firth, Y. Shimoda, A cross analysis of existing methods for modelling household appliance use. J. Build. Perform. Simul. **12**(2), 160–179 (2019). https://doi.org/10.1080/19401493.2018.1497087
70. S. Yilmaz, S.K. Firth, D. Allinson, Occupant behaviour modelling in domestic buildings: the case of household electrical appliances. J. Build. Perform. Simul. **10**(5–6), 582–600 (2017)

Chapter 2
Probabilistic Model and Prediction of Vehicle Daily Use

Shinkichi Inagaki and Tatsuya Suzuki

2.1 How to Use a Vehicle at Home

In this chapter, prediction methods for a future timing profile of when an electric vehicle (EV) is parked at a home or in transit away from the home are introduced. Such a profile plays an important role in a home energy management system (HEMS), which may utilize in-vehicle battery of EVs. The HEMS can charge and discharge the battery only while the vehicle is connected to the charging/discharging station located at the home (Figs. 2.1 and 2.2). This means that the HEMS is only able to utilize the energy stored in the battery for electric consumption in the home while the vehicle remains parked at home. Of course, because the prime objective of the vehicle is transportation, HEMS needs to utilize the stored electricity without interfering with this objective as much as possible. For example, HEMS has to store enough energy in its battery by the time the vehicle leaves home so the owner can use the vehicle for transportation. In order to plan charging and discharging the in-vehicle battery at home, the HEMS needs to know the future profile regarding when the vehicle will be available at home.

One way to achieve the future usage profile of the vehicle in terms of traveling or parking is based on the user's schedule. If the user writes her/his plan for use of the vehicle in a schedule, and the HEMS is allowed to access this, a specific future profile is in turn, available. When the user changes her/his plans for use of the vehicle and updates the schedule, the EMS should ideally also access the updated future profile.

The other way is prediction. The prediction of such future profile is based on data obtained from the daily driving routine. Such data is possible to gain by using a GPS

S. Inagaki (✉) · T. Suzuki
Department of Mechanical Systems Engineering, Nagoya University, Nagoya, Aichi, Japan
e-mail: inagaki@nagoya-u.jp

T. Suzuki et al. (eds.), *Design and Analysis of Distributed Energy Management Systems*, Power Electronics and Power Systems,
https://doi.org/10.1007/978-3-030-33672-1_2

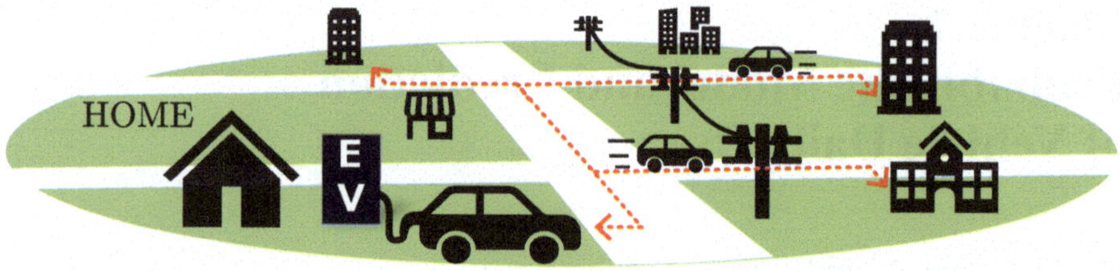

Fig. 2.1 A car is used in an daily life

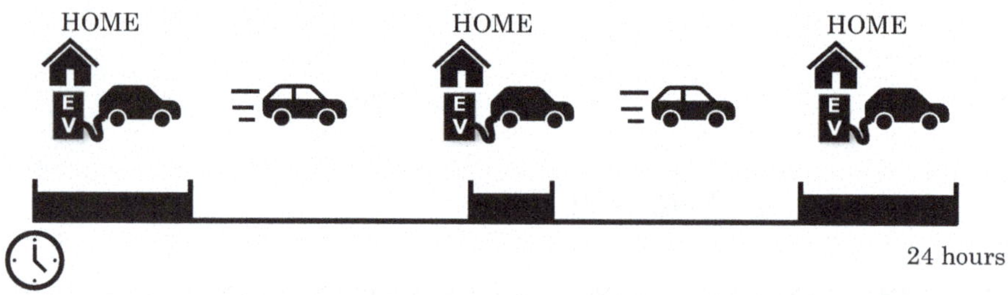

Fig. 2.2 Profile of vehicle parked at home or traveling; the vehicle is charged/discharged at home

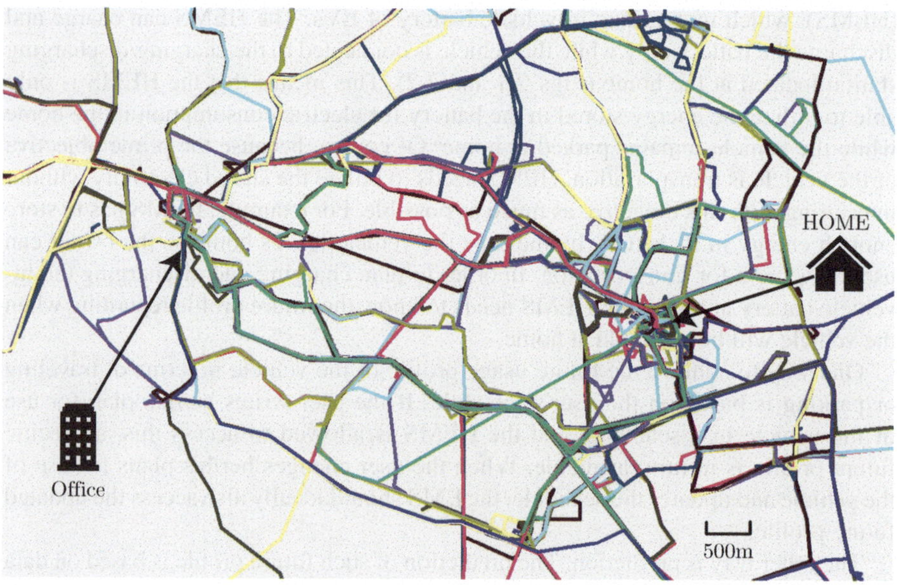

Fig. 2.3 Tracks in a GPS logger installed in vehicle used in a daily-basis. Roads are not displayed and only tracks are shown. Data shown was recorded for about 1 year and 2 months

logger for example. Figure 2.3 shows tracks for about 1 year and 2 months, which are stored in a GPS logger installed in a vehicle owned by a male user. He drives the vehicle mainly for commuting between his home and office. While many routes and other destinations are also observed, some regularities are found in the departure time from home and the arrival time at home. The GPS logger started recording at the same time as the vehicle started running, and stopped when the vehicle was parked.

When the prediction of the future profile is updated, it should reflect the present state of the vehicle, parked or traveling. If the prediction only depends on the daily driving data, then it only establishes a fixed pattern for the future profile. The initial state of the future profile should be the same as the present state of the vehicle. By using an in-home communication network, the HEMS can acquire information about whether or not the vehicle is currently connected to the charging/discharging station. The prediction methods introduced in this chapter are based on two types of data, statistical data, which are obtained from the user's daily use of the vehicle, and observation data, which is obtained by observing whether the current state of the vehicle is parked or not.

2.2 Related Works

Predicting a vehicle's departure and arrival times is itself useful for our daily life. For example, a precise prediction of arrival time of a vehicle is required for car navigation systems and electric maps [2, 18, 24]. In these studies, departure time is given and arrival time is predicted, considering the time zone and the traffic conditions. Etterna et al. [6, 7] have estimated the profile of departure and arrival times for a vehicle fleet based on a trip and activity utility model. Prediction of the arrival time of a bus using GPS data is studied by Lin et al. [16].

If charging/discharging of EVs is controlled properly, it is expected to contribute to stabilizing the electric quality in the electricity grid [12, 13, 23]. On the other hand, one concern is that uncontrolled vehicle fleet charging behavior may cause an impact, having mainly negative effects on the grid. Supposing a growing penetration of EVs, the future impact is analyzed based on statistical data about people's charging behavior. Shahidinejad et al. [21] and Ashtari et al. [1] investigated the electricity demand to charge a vehicle fleet in a city, based on GPS data recorded in Winnipeg. Darabi and Ferdowsi [4] estimated charging load profiles according to the class of car and the means of charging.

Estimation of an individual driver's charging behavior was studied on the basis of analyzing her/his driving behavior [15]. This study aimed to estimate the charging time and the amount of charging electricity. Grahn et al. [8] modeled the driver's charging behavior correlating with the behavior of her/his house. Kessler and Bogenberger [14] analyzed driving and charging behaviors based on data obtained from 40 real electric vehicles in Germany. Grahn et al. [9] modeled the charging behavior, considering the movement between multiple locations, various purposes to use the vehicle, and the electricity price.

The profile of departure and arrival time of an EV is important for planning charging/discharging of the in-vehicle battery, as mentioned in the previous section.

In [19], the charging profile is derived, to minimize the electricity cost, while departure and arrival times, driving behavior, and charging time are all given. In [3, 22], the time slot to charge the EV is given, and the charging profiles of the EV fleet are derived. Pantos [17] also calculated charging profiles of an EV fleet based on optimization. In this study, the profiles of departure and arrival times are also given, but they are obtained from a statistical data. de Hoog et al. [5] addresses an optimization problem to obtain charging/discharging profiles for EVs. At each time an EV arrives at a charging/discharging station, its next departure time is sent to the planner. The planner calculates the charging/discharging profile, so that the in-vehicle battery is fully charged by the departure time.

Iversen et al. [11] propose an architecture to optimize the profile of charging/discharging based on prediction of vehicle use. The vehicle use is modeled using a semi-Markov model, which is a type of Markov model whose transition probability changes with the passage of time. Yamaguchi et al. [26] proposed a real-time prediction algorithm of departure and arrival times of a vehicle based on greedy algorithm. In [25], they proposed another algorithm based on the Markov model and dynamic programming. Ito et al. [10] implemented the prediction algorithm [25] in a home energy management system which utilizes the in-vehicle battery in the house, and conducted an experiment in a real system. Hereafter, this chapter explains the real-time prediction algorithm of departure and arrival times of a vehicle based on the Markov model and dynamic programming [20].

2.3 Statistical Data of Daily Car Use

Departure time from home and arrival time at home can be extracted from the GPS logger, based on the latitude and longitude and the recorded time. When the vehicle departs from the latitude and longitude corresponding to the home, the departure time is recorded, and when returns, the arrival one is also registered. The recorded time is supposed to be a discrete value with 30 min increments, as shown in Fig. 2.4. This period depends on the frequency of the planned HEMS charging and discharging scheme. If HEMS requires a higher frequency, the time period should decrement accordingly. In case (a), the vehicle departs home just after 7:00 and arrives at home just before 7:30. In this case 7:00 is counted as the departure time and the arrival time is 7:30. Similarly, case (b) is counted as (departure time, arrival time) = (7:00, 8:00), case (c) is (7:00, 8:30), and (d) is (7:00, 9:00). Thus, each journey time (travel time) is assumed to take 30 min. This is a somewhat rough assumption because, even when the travel time is 1 min, as in case (a), it is counted as 30 min. If the traveling time crosses 7:30, as in case (b), it is also counted as 1 h total travel time. To avoid this, short journey times are omitted from the data. Such short journey time is unusual in many cases, and only a little amount of energy in the in-vehicle battery is consumed.

Tables 2.1 and 2.2 are examples of the frequency distribution regarding the vehicle use. The column in Table 2.1 represents the departure time from home and

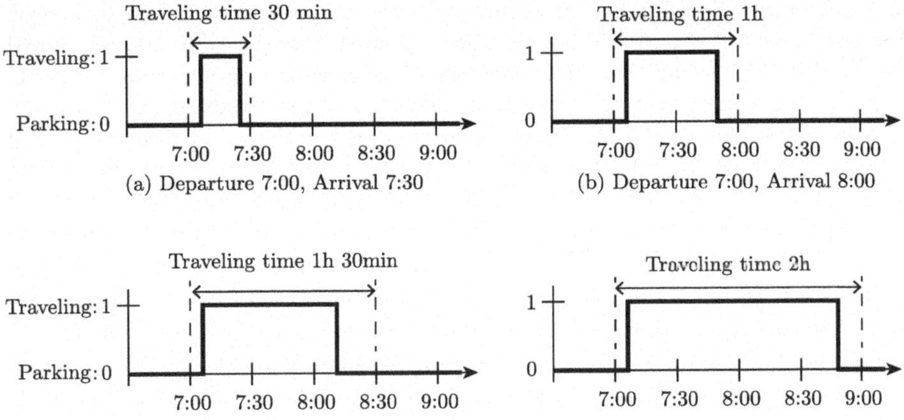

(a) Departure 7:00, Arrival 7:30 (b) Departure 7:00, Arrival 8:00

(c) Departure 7:00, Arrival 8:30 (d) Departure 7:00, Arrival 9:00

Fig. 2.4 Departure and arrival time

Table 2.1 Example of frequency distribution of vehicle traveling times

Duration time		Departure time (time index) 0:00 (0)	0:30 (1)	1:00 (2)	1:30 (3)	...	23:00 ($T-2$)	23:30 ($T-1$)
0:30	(1)	3	1	2	4	...	1	0
1:00	(2)	2	1	1	6	...	0	1
1:30	(3)	4	2	0	7	...	1	2
⋮		⋮	⋮	⋮	⋮	⋱	⋮	⋮
23:00	($T-2$)	2	4	0	4	...	3	1
23:30	($T-1$)	1	3	1	2	...	1	2
24:00	(T)	0	1	0	1	...	1	0

The notation (departure time, duration time) = (1:00, 0:30) means that the vehicle departs from home at 1:00 and arrives home at 1:30

Table 2.2 Example of frequency distribution of vehicle parking times

Duration time		Parking start time (time index) 0:00 (0)	0:30 (1)	1:00 (2)	1:30 (3)	...	23:00 ($T-2$)	23:30 ($T-1$)
0:30	(1)	3	1	2	4	...	1	0
1:00	(2)	2	1	1	6	...	0	1
1:30	(3)	4	2	0	7	...	1	2
⋮		⋮	⋮	⋮	⋮	⋱	⋮	⋮
23:00	($T-2$)	2	4	0	4	...	3	1
23:30	($T-1$)	1	3	1	2	...	1	2
24:00	(T)	0	1	0	1	...	1	0

The notation (arrival time, duration time) = (2:00, 0:30) means that the vehicle arrives home at 2:00 and departs from home at 2:30

the column in Table 2.2 the arrival time at home. The row of the tables represents the corresponding duration time. Each set extracted from the GPS data is counted in the cell. The former is a frequency distribution about the departure time and the corresponding travel time, which is called SDTT (Statistics of Departure and Travel Time). The latter is a frequency distribution about the arrival time and parked time, which is called SAPT (Statistics of Arrival and Parked Time). The travel time is obtained from the duration of time between the departure time and the arrival time. The parked time is the duration time between the arrival time and the next departure time. The rows in these tables are duration times of traveling and parking, respectively. Duration times of longer than 24 h are omitted because almost all daily used vehicles are expected to return home the same day. If a vehicle frequently travels for multiple days, the rows of the table should be extended. In this chapter, a method to predict the future profile of the vehicle's departure and arrival time using these frequency distributions is introduced.

2.4 Preparation for Problem Formulation

2.4.1 Description of Time

As shown in Fig. 2.5, time is digitized as (0:00, 0:30, 1:00, ..., 23:00, 23:30). To simplify the description, the digitized time is indexed as $\mathbf{T} = \{0, 1, 2, \ldots, 46, 47\ (= T-1)\}$ in ascending order of powers. The maximum number of the time index T is calculated by

$$T = 24:00/\Delta t = 48, \tag{2.1}$$

where Δt is 30 min. By selecting Δt so that $24:00/\Delta t$ is an integer number, the indexed time becomes cyclic, $\mathbf{T} \subset C_T$. C_T is a cyclic group whose order is T. When $a, b \in C_T$,

$$a + b \equiv (a + b) \pmod{T}. \tag{2.2}$$

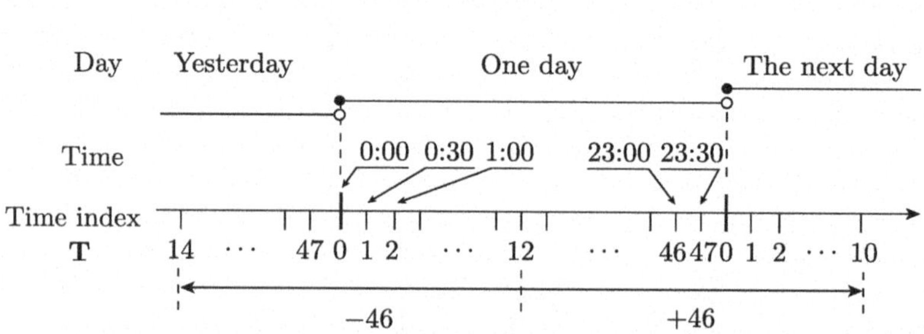

Fig. 2.5 Time and time indices with $T = 48$: time index is cyclic in each day. For example, $12 + 46 \equiv (12 + 46) \pmod{48} = 10$, and $12 - 46 \equiv (12 - 46) \pmod{48} = 14$

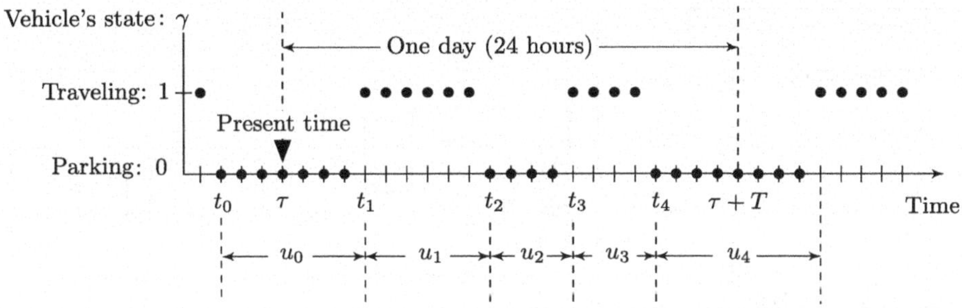

Fig. 2.6 Example of a profile of departure and travel time (PDTT) for 24 h (T) starting at present time. Two states are used to describe the vehicle, traveling ($\gamma = 1$) and parking ($\gamma = 0$), depicted by the black dots. The state is assumed to be sampled at a discrete interval, for example, 30 min

For example, when $T = 48$ and $12,\ 46 \in \mathbf{T}$

$$12 + 46 \equiv (12 + 46) \pmod{48} = 10 \in \mathbf{T}. \tag{2.3}$$

This means that the time index of 46 steps ahead from 12 is 10 of the next day. In addition,

$$12 - 46 \equiv (12 - 46) \pmod{48} = 14 \in \mathbf{T}. \tag{2.4}$$

In this case, the time index of 46 steps before from 12 is 14 of yesterday.

The set of indices for departure time is defined as $\mathbf{T}$. The set of duration time is defined as $\mathbf{D} = \{1, 2, 3, \ldots, 47, 48\ (= T)\}$ as shown in Tables 2.1 and 2.2.

2.4.2 *Expression of Future Profile of Vehicle Use*

The availability of a vehicle at home is expressed by two states, parked at home and away from home. The profile of the two states for 24 h from the present time is called PDTT (Profile of Departure and Travel Time). Figure 2.6 is an example of a PDTT, which is expressed by sets of start parking time and its duration , or the sets of start traveling time and its duration. The present time is $\tau \in \mathbf{T}$ and the time value for the 24 h ahead is $\tau + T$. The ith start time of traveling or parking is denoted as $t_i \in \mathbf{T}$ and its duration time is $u_i \in \mathbf{D}$. In Fig. 2.6, the vehicle departs from home at t_1 and t_3, and arrives at home at t_2 and t_4. The vehicle is, for example, away from home for u_1 from t_1 to t_2 and parked for u_2 from t_2 to t_3.

2.5 Problem Formulation of Predicting a PDTT

In this section, an estimation problem of a PDTT is formulated based on a maximum-likelihood problem, where a most probable profile of parking and traveling for 24 h is derived. The EMS is assumed to be able to observe whether

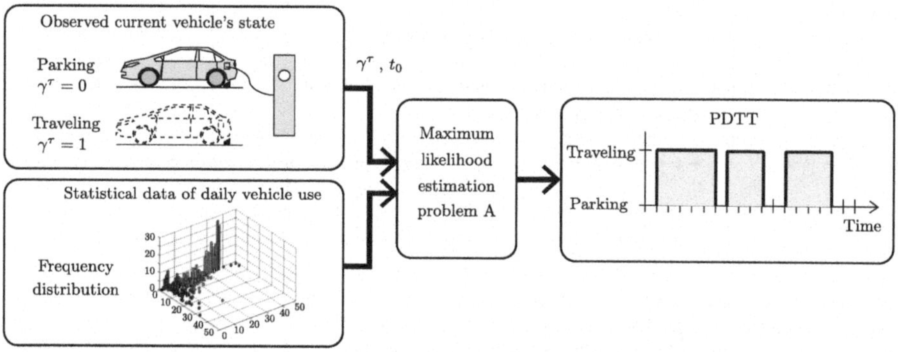

Fig. 2.7 Procedure to derive PDTT, based on the maximum-likelihood estimation problem

or not the vehicle is parked at home. The observed current state of the vehicle is expressed as $\gamma^\tau \in \{0, 1\}$, where 0 means that the vehicle is parked at home and 1 that the vehicle is away from home. In addition, EMS is assumed to know the latest start time of the vehicle parking or traveling, and which is represented as $t_0 \in \mathbf{T}$. Figure 2.7 displays the procedure to predict a certain PDTT. First, EMS collects the two parameters of the current vehicle situation, γ and t_0, and refers to the statistical data of the daily vehicle use, SDTT&SAPT. Then the prediction problem of a PDTT is formulated as a maximum-likelihood estimation problem as follows:

A. Maximum likelihood estimation of the PDTT at time τ

Given

$$(\gamma^\tau, t_0) \in \{0, 1\} \times \mathbf{T},\ \text{SDTT\&SAPT};$$

find

$$u_0 \in \mathbf{D},\ \{(t_i, u_i) \in \mathbf{T} \times \mathbf{D} \mid i = 1, \ldots, m\},\ m \in \mathbb{Z}_+;$$

which maximize

$$J = \Pr(U_0 = u_0,\ (T_1, U_1) = (t_1, u_1),\ \ldots,\ (T_m, U_m) = (t_m, u_m)); \tag{2.5}$$

subject to

$$t_i = t_{i-1} + u_{i-1}, \quad \forall i \in \{1, 2, \ldots, m\}, \tag{2.6}$$

$$t_0 + u_0 > \tau, \tag{2.7}$$

$$t_m + u_m > \tau + T. \tag{2.8}$$

Part of the variables used in this formulation is shown in Fig. 2.6. The ith start time of parking or traveling is $t_i \in \mathbf{T}$, and T_i is its stochastic variable. $u_i \in \mathbf{D}$ is the duration time of ith parking or traveling, and U_i is its stochastic variable. u_0 is the duration time of the latest parking or traveling started at t_0. u_0 is also a decision variable. This means that the EMS does not know the duration time u_0 of the latest parking and traveling. In other words, the latest parking or traveling is continuing at τ from t_0. $m \in \mathbb{Z}_+$ is the number of times that the vehicle's parking and traveling are initiated between τ to $\tau + T$. The variable m is also a decision variable in this optimization problem. In the case of Fig. 2.6, m is 4.

Equation (2.6) represents the relationship between t_i and t_{i+1}. Equation (2.7) means that the latest parking or traveling is continuing at the present time τ. Equation (2.8) means the total number of times of parking and traveling between τ and $\tau + T$ is m.

Evaluation function (2.5) is a joint probability function from the combination of $2m + 1$ discrete variables. The number of all possible patterns of PDTT is up to $2^T = 2^{48} \approx 281 \times 10^{12}$. It is difficult to find an optimal variable in a realistic time by directly calculating the joint probability. In the following sections, a method to find an approximate solution to this problem in a more realistic time frame is introduced.

Evaluation function (2.5) is a joint probability of a combination of $2m+1$ discrete variables, in addition m is also a decision variable. The number of all possible patterns of PDTT is up to $2^T = 2^{48} \approx 281 \times 10^{12}$. It is difficult to find an optimal variables in realistic time frame by directory directly calculating the joint probability. In the following sections, a method to find an approximate solution of this problem in short time is introduced.

2.6 Markov Model Representing Production Process of a PDTT

2.6.1 *A PDTT and a Markov Model*

A PDTT is composed of the starting time of traveling or parking and its duration time, as shown in Fig. 2.6. It is possible to correspond the state of each time in a PDTT to the remaining time when the vehicle state continues. Figure 2.8 is a profile of the remaining time of a vehicle state, which is converted from Fig. 2.6. The vertical axis shows the remaining time when the vehicle state changes to the next one. Then, the remaining time decreases with time passing and the vehicle state changes to the next one when the remaining time runs out. This procedure can be represented by a stochastic model, such as a Markov model.

A Markov model expresses a process of state changes. The model is composed of three components, a set of states, a set of transition probabilities, and a set of initial probabilities. An initial state is determined based on the initial probability. The next state to which a state transits is based on the transition probability. The Markov model, which represents the procedure to create a PDTT in Fig. 2.6 or 2.8, is defined as follows:

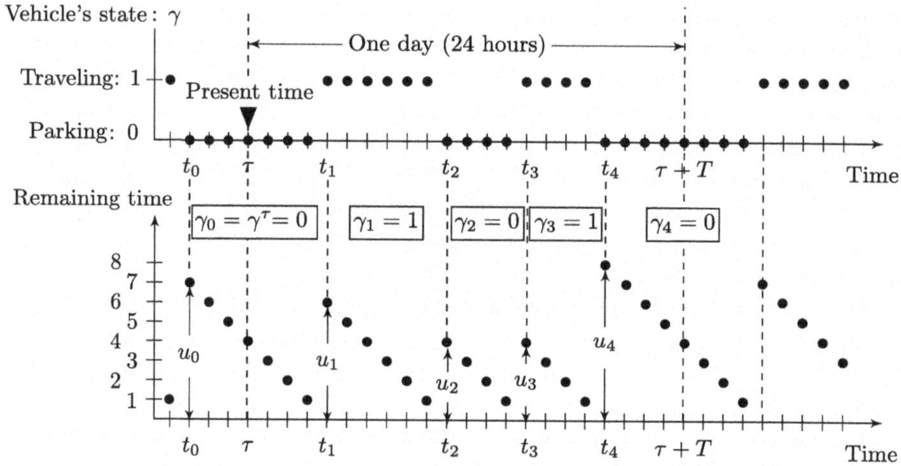

Fig. 2.8 Conversion from a PDTT to a profile of remaining time of a vehicle's state: the vehicle's state changes when the remaining time runs out

$$\mathscr{M}^{\tau} = (\mathbf{S}, A, \Pi^{\tau}) \tag{2.9}$$

$\mathbf{S}^{\tau} = \{0, 1\} \times \mathbf{D} \times \mathbf{T}$ is the set of states, $A = \{a_{ss'} \in [0, 1] \mid (s, s') \in \mathbf{S} \times \mathbf{S}\}$ is the set of transition probabilities, and $\Pi^{\tau} = \{\pi_s \in [0, 1] \mid s \in \mathbf{S}\}$ is the set of initial probabilities. Π^{τ} changes with the present time τ because it is calculated based on the observation of the vehicle's state. Therefore, the Markov model consequently depends on the present time τ.

The Markov model $\mathscr{M}^{\tau}$ is shown in Fig. 2.9 as a left-to-right model. The state is drawn as a white circle. The upper half of the states are the ones for the vehicle's state when traveling, $\gamma = 1$, and the lower half are for parking, $\gamma = 0$. The column means the time index in $\mathbf{T}$ and the row means the remaining time in $\mathbf{D}$. The state transition is represented by an arrow connected from one state to another. The color of state transitions is partially changed for viewability. The states in the left edge are the initial states.

If a PDTT is assumed to be known as Fig. 2.8, the sequence of state transitions can be drawn as shown as in Fig. 2.10. Each vehicle's state γ_i starts at t_i, with possessing the having a remaining time u_i. The remaining time decreases by 1 ($= \Delta t$) with time passinges. When the remaining time runs out, the next vehicle's state changes to γ_{i+1}. Evidently, such the future transition sequence is not known at the present time τ. In addition, while the latest time of the vehicle's state t_0 and γ^{τ} is known, its remaining time u_0 is unknown at τ. This unknown information is calculated based on probability.

2.6.2 *Calculation of Probabilities*

A state in which the vehicle's state $\gamma \in \{0, 1\}$ has a remaining time $u \in \mathbf{D}$ at time $t \in \mathbf{T}$ is represented as (γ, u, t). The transition probability from a state (γ, u, t)

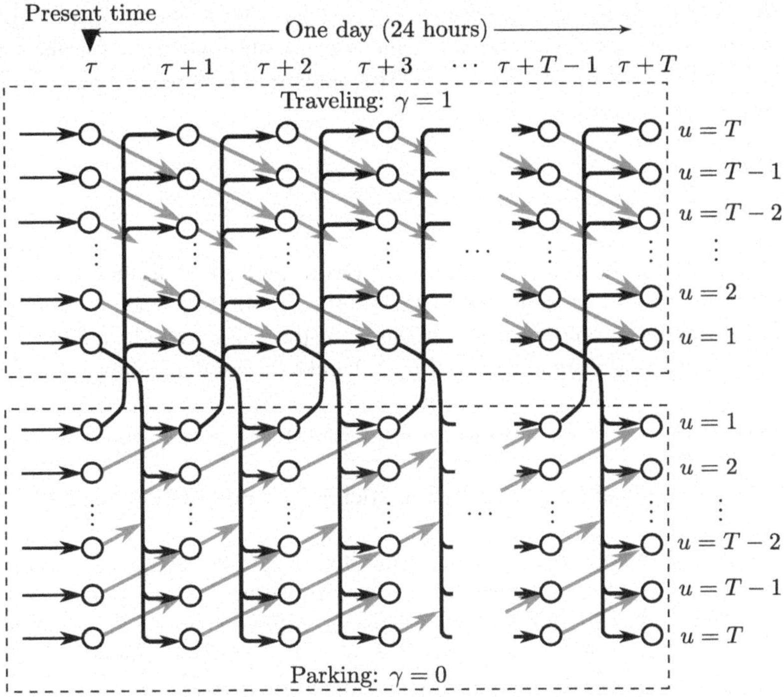

Fig. 2.9 Left-to-right model of PDTT generation: state transitions from the states $u = 1$, which is drawn by a black curved arrow, are partially drawn for the sake of clarity

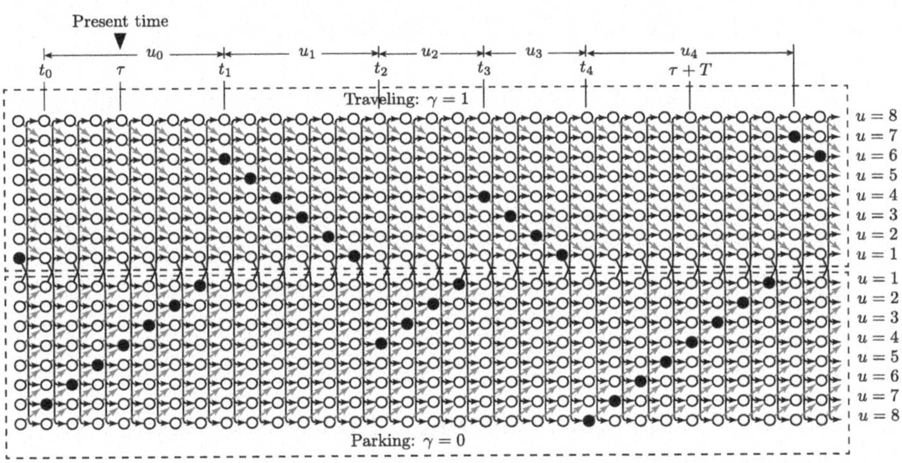

Fig. 2.10 A sequence of state transitions if the PDTT is produced as Fig. 2.8; the black circles are the existing states. Note that such a future sequence is not known at the present time, but derived based on probabilistic calculation

to a state (γ', u', t') is represented as $a_{(\gamma,u,t)(\gamma',u',t')}$. First, the state transition by which the value of vehicle's state γ changes occurs only when the former state of the transition has remaining time $u = 1$. This transition is drawn by a curved black arrow in Fig. 2.9 and expressed as follows:

$$a_{(1-\gamma,1,t-1)(\gamma,u',t)} = \frac{c_\gamma(t, u')}{\sum_{\delta \in \mathbf{D}} c_\gamma(t, \delta)}, \tag{2.10}$$

where $c_\gamma(t, u')$ is the value of cell of uth row and tth column of the SDTT (Table 2.1) when $\gamma = 1$, and one of the SAPT (Table 2.2) when $\gamma = 0$. Next, the remaining time monotonically decreases in the same vehicle's state γ. This transition is drawn by a gray arrow in Fig. 2.9 and expressed as follows:

$$a_{(\gamma,u+1,t-1)(\gamma,u,t)} = 1. \tag{2.11}$$

Other transition probabilities which are not defined by Eqs. (2.10) and (2.11) are all 0.

The initial probabilities are calculated based on the observed information (t_0, γ^τ). When the observed vehicle's state is traveling, $\gamma^\tau = 1$, then t_0 is the latest departure time. Figure 2.11 shows the possible states which are drawn by a black circle. The states that possibly exist at the present time τ are ones of $u \in \{1, 2, \ldots, T - (\tau - t_0) \pmod T\}$. These states are transited from the states at t_0 of $u \in \{(\tau - t_0) \pmod T) + 1, (\tau - t_0) \pmod T)] + 2, \ldots, T\} = \mathbf{D}'$. When $\gamma^\tau = 0$, t_0 is the latest arrival time (the latest time when the vehicle initiated parking). In this case, Fig. 2.12 shows the possible states. In two cases, the initial probabilities at the present time τ are calculated based on the transition probabilities from $t_0 - 1$ to t_0:

$$\pi_{(\gamma^\tau,u,\tau)} = \begin{cases} \dfrac{a_{(1-\gamma^\tau,1,t_0-1)(\gamma^\tau,u,t_0)}}{\sum\limits_{\delta \in \mathbf{D}'} a_{(1-\gamma^\tau,1,t_0-1)(\gamma^\tau,\delta,t_0)}} & \text{if } u \in \{1, 2, \ldots, T - (\tau - t_0) \pmod T)\}, \\ 0 & \text{otherwise.} \end{cases} \tag{2.12}$$

$$\pi_{(1-\gamma^\tau,u,\tau)} = 0. \tag{2.13}$$

2.7 Solving Method for Predicting a PDTT by Dynamic Programming

As noted in Sect. 2.5, the original formulation of a maximum-likelihood problem to produce a most plausible PDTT is highly computational demanding. It is not realistic to calculate the original problem A. Thus, the procedure to derive a PDTT

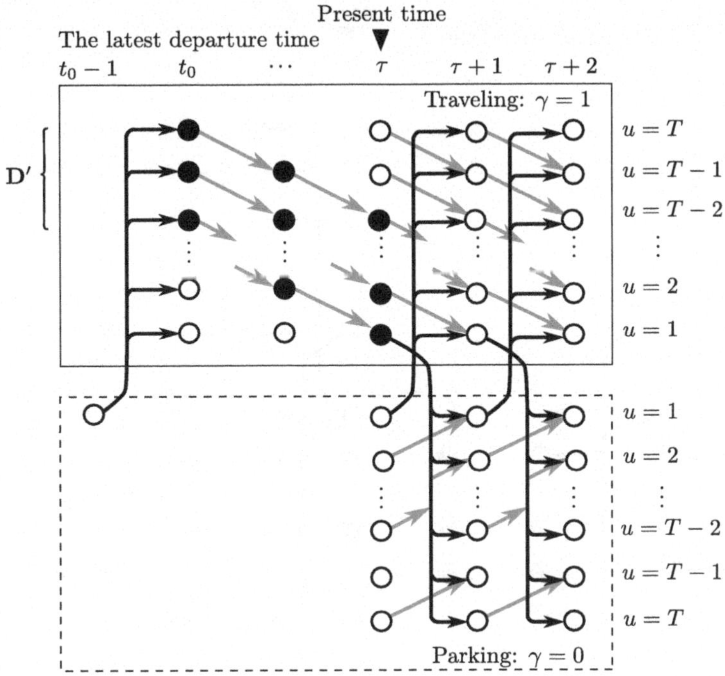

Fig. 2.11 Possible initial states in the case of $\gamma^\tau = 1$ at the present time τ, which are rooted the states transited at time t_0

is modified from Fig. 2.7 to Fig. 2.13. First, the Markov model $\mathscr{M}^\tau$, Eq. (2.9), is produced from the observed current vehicle's state and the SDTT&SPAT, as explained in Sect. 2.6. Then, an approximated formulation of the problem A is calculated based on dynamic programming. The approximated formulation is described by the following most-likelihood problem:

A'. Maximum likelihood estimation of the PDTT on a Markov model $\mathscr{M}^\tau$

Given: $\mathscr{M}^\tau$

find: $\{s(t) \in \mathbf{S} \mid t \in \mathbf{T}\}$

which maximize:

$$\begin{aligned} J &= \Pr(S(\tau) = s(\tau), S(\tau+1) = s(\tau+1), \ldots, S(\tau+T) = s(\tau+T)) \\ &= \Pr(S(\tau) = s(\tau)) \Pr(S(\tau+1) = s(\tau+1) \mid S(\tau) = s(\tau)) \cdots \\ &\qquad \Pr(S(\tau+T) = s(\tau+T) \mid S(\tau+T-1) = s(\tau+T-1)) \\ &= \pi_{s(\tau)} a_{s(\tau)s(\tau+1)} \cdots a_{s(\tau+T-1)s(\tau+T)} \end{aligned} \tag{2.14}$$

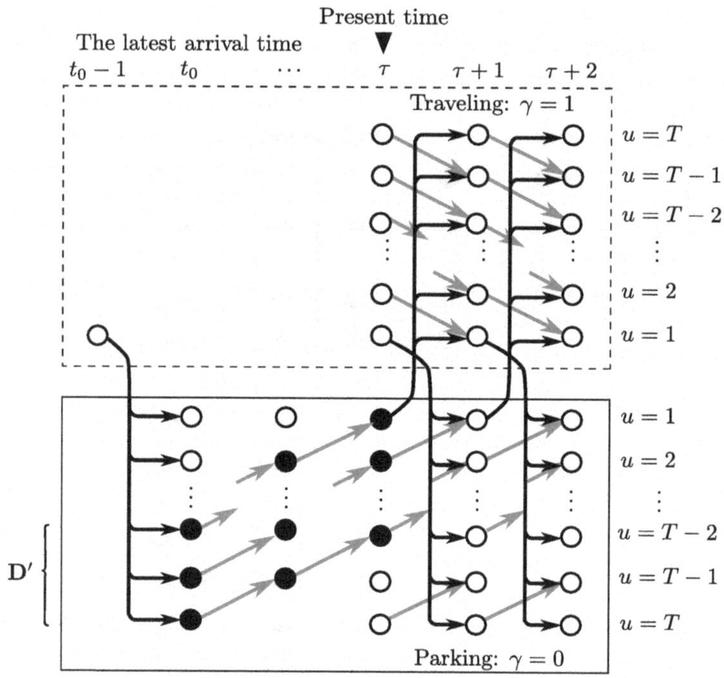

Fig. 2.12 Possible initial states in the case of $\gamma^\tau = 0$ at the present time τ, which are rooted the states transited at time t_0

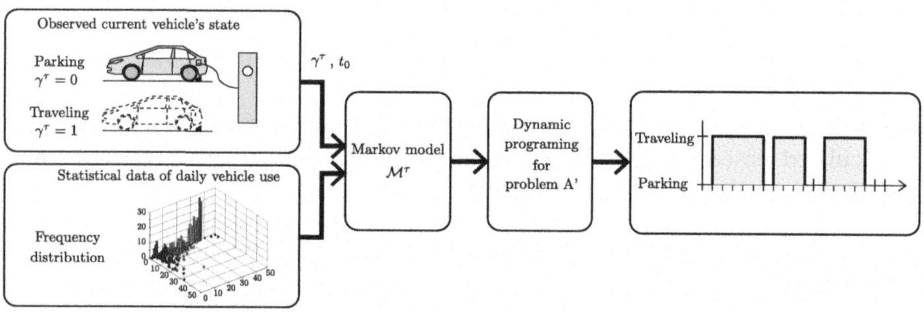

Fig. 2.13 Modified procedure to derive PDTT based on the Markov model and dynamic programming

where a state (γ, u, t) is simplified as $s(t)$, and $S(t)$ is the stochastic variable of $s(t)$. The evaluation function J, (2.14), is the likelihood to produce a PDTT by the Markov model. This equation is represented as the product of the transition probabilities. Therefore, it is possible to calculate the maximization of J by dividing the problem into sub-problems at each time. The algorithm for this maximization is described by dynamic programming.

A variable $\delta_{s(\tau)}$ is defined as follows:

$$\delta_{s(t)} = \begin{cases} \pi_{s(\tau)} & \text{if } t = \tau, \\ \max\limits_{s(t-1)} \{\delta_{s(t-1)} a_{s(t-1)s(t)}\} & \text{otherwise.} \end{cases} \tag{2.15}$$

This variable can be sequentially derived by forward calculation from $t = \tau$. By using this variable, the maximum value of the evaluation function J and the estimated state at the last time $t + T$ are derived as follows:

$$J^* = \max_{s(\tau+T)} \delta_{s(\tau+T)}, \tag{2.16}$$

$$s^*(t+T) = \arg\max_{s(\tau+T)} \delta_{s(\tau+T)}. \tag{2.17}$$

Then, the optimal state transition profile $s^*(\tau), \ldots, s^*(\tau+T)$ is recursively obtained by backward calculations:

$$s^*(t) = \arg\max_{s(t)} \{\delta_{s(t)} a_{s(t)s^*(t+1)}\}$$
$$\text{if } t \in \mathbf{T}^\tau \setminus \{\tau + T\}. \tag{2.18}$$

2.8 Simulation Result of the PDTT Estimation

The estimation algorithm introduced in Sect. 2.7 is verified using vehicles data obtained from real-scenarios. The sets of frequency distributions, SDTT&SAPT, were obtained from two vehicles. A GPS logger installed in each vehicle stored data of daily vehicle usage for 27 months. The SDTT&SAPT distributions were made using data stored for 24 months. An additional 3 months-dataset was used for verification. Simulation was performed under a situation in which (t_0, γ^τ) was observed from the data verified every 30 min. A PDTT was produced, based on calculating the dynamic programming, through Eqs. (2.15)–(2.18), for deriving a solution to problem A' in Sect. 2.7 in each 30 min. The simulation was conducted using an Intel(R) Core^{TM} i7-4770 CPU 3.40 GHz and RAM 8 GB personal computer.

Vehicle A was used by a middle-aged woman with the main purpose of commuting and shopping. Vehicle B was used by a young woman for mainly shopping, picking up her child, and leisure purposes. The obtained frequency distribution of SDTT&SAPT is shown in Fig. 2.14. The three-dimensional box is composed of three axes: start traveling or parking time, its duration and the frequency at which the combination of start, and duration time occurs. The peak in distributions of vehicle A is higher that of vehicle B. This means vehicle A was used in a more regular manner than vehicle B.

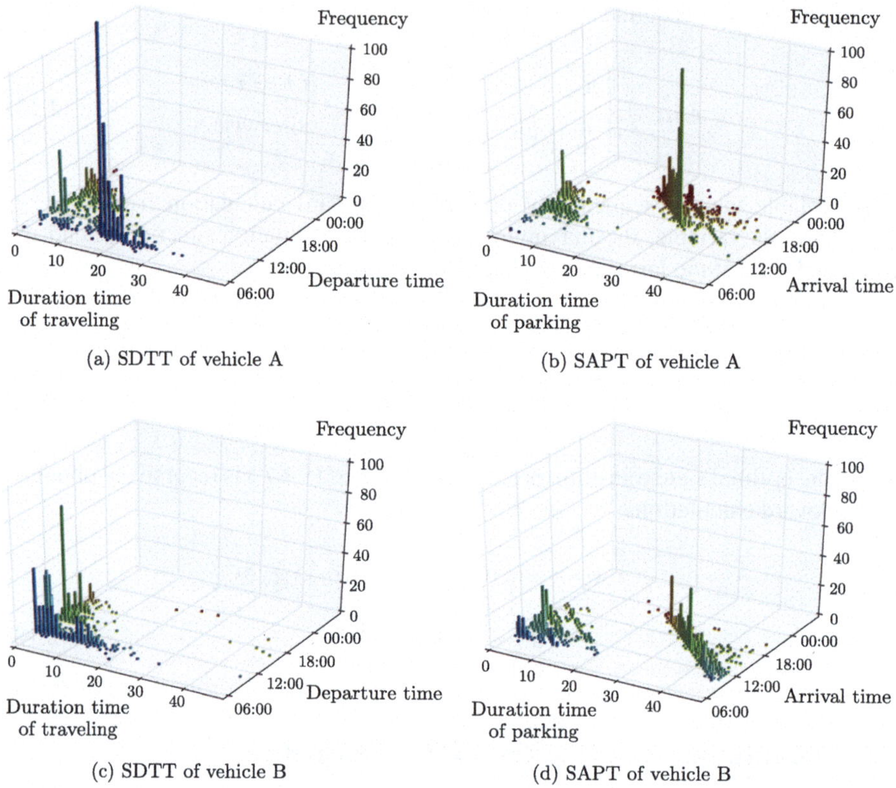

Fig. 2.14 Frequency distribution SDTT&SAPT of two vehicles, A and B (Reprinted with permission from [20])

The estimated patterns of PDTT are shown in Fig. 2.15. The solid line is the ground truth of future PDTT. The dashed line is the estimated pattern of PDTT. Each pattern was produced every 30 min, from top to bottom. Thus, the present time on the left side of the horizontal axis also increases by every 30 min. The computational time to derive each pattern was 37.7 ms on average. The ground truth of future PDTT shifts from right to left with every 30 min, without changing the shape of the pattern. The estimated pattern of PDTT also shifts from right to left without changing its shape when the estimated vehicle state at the present time is the same as the ground truth. On the other hand, when the estimated vehicle's state at the present time is different to the ground truth, the pattern of estimation changes, so as to match the present vehicle state. This is because the observed vehicle state (t_0, γ^{τ}) is changed and the initial probabilities (2.12) and (2.13).

The performance of prediction is calculated based on accuracy and recall. Accuracy is the agreement rate between the predicted profile and the ground truth. Recall is the rate of the correct answer when predicting that the vehicle is absent from home (see Sect. 9.6.2). Accuracy and recall should be high for the EMS.

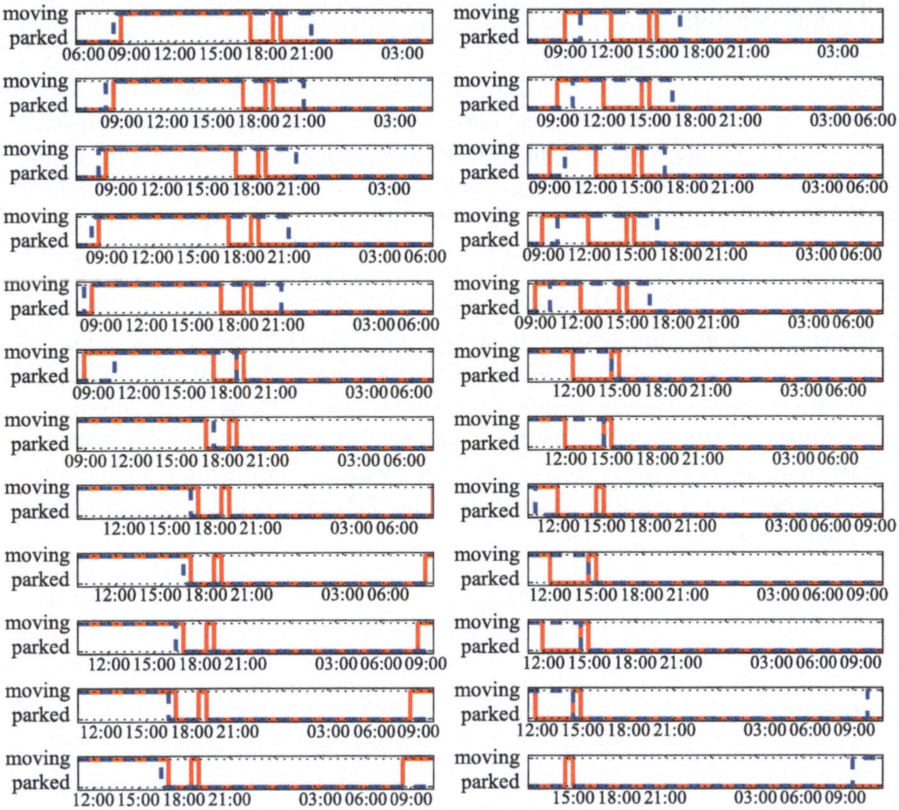

Fig. 2.15 Estimated PDTT every 30 min: the left side of the horizontal axis is the present time. The solid line is the ground truth of future PDTT and the dashed line is the estimated one (Reprinted with permission from [20])

Table 2.3 Accuracy and recall of vehicles A and B

	Accuracy	Recall
Vehicle A	72%	64%
Vehicle B	73%	56%

Especially, recall is essential for stable operation of the EMS, because the EMS is likely to plan charging/discharging while the vehicle is absent when recall is low. The average accuracy and recall are shown in Table 2.3. The accuracies of two vehicles are almost same, but recall of vehicle A is higher than one of B. This is because vehicle A was used in a more regular manner than vehicle B.

2.9 Conclusion

In this chapter, the real-time prediction algorithm of departure and arrival times of a vehicle based on the Markov model and dynamic programming is introduced. To utilize the in-vehicle battery in an EMS, considering the acceptance by the users, the EMS should be able to identify when the vehicle is being driven and when it is parked, which represents the profile of departure and travel time (PDTT). Daily vehicle usage data is recorded, and a frequency distribution about the departure time and the corresponding travel time (SDTT) and a frequency distribution about the arrival time and parked time (SAPT) are registered. A PDTT over 1 day is calculated every 30 min based on the PDTT, the SDTT, and the present state of the vehicle, parked or traveling. Simulations which were performed for two vehicles showed accuracy of approximately 70% and recall of 54–64%. To improve the prediction performance, it is effective to utilize a calender of users and other online information, such as traffic information and weather information. Prediction of departure and arrival times of a vehicle that runs between multiple stations is one of our future works.

Acknowledgement This research was supported by JST CREST Grant NumberJPMJCR15K3.

References

1. A. Ashtari, E. Bibeau, S. Shahidinejad, T. Molinski, PEV charging profile prediction and analysis based on vehicle usage data. IEEE Trans. Smart Grid **3**(1), 341–350 (2012)
2. M. Chen, S.I.J. Chien, Dynamic freeway travel-time prediction with probe vehicle data: link based versus path based. Transp. Res. Record J. Transp. Res. Board **1768**, 157–161 (2001)
3. K. Clement-Nyns, E. Haesen, J. Driesen, The impact of charging plug-in hybrid electric vehicles on a residential distribution grid. IEEE Trans. Power Syst. **25**(1), 371–380 (2010)
4. Z. Darabi, M. Ferdowsi, Aggregated impact of plug-in hybrid electric vehicles on electricity demand profile. IEEE Trans. Sustain. Energy **2**, 501–508 (2011)
5. J. de Hoog, T. Alpcan, M. Brazil, D.A. Thomas, I. Mareels, Optimal charging of electric vehicles taking distribution network constraints into account. IEEE Trans. Power Syst. **30**(1), 365–375 (2015)
6. D. Ettema, H. Timmermans, Modeling departure time choice in the context of activity scheduling behavior. Transp. Res. Rec. J. Transp. Res. Board **1831**, 39–46 (2003)
7. D. Ettema, F. Bastin, J. Polak, O. Ashiru, Modelling the joint choice of activity timing and duration. Transp. Res. A Policy Pract. **41**(9), 827–841 (2007)
8. P. Grahn, J. Munkhammar, J. Widen, K. Alvehag, L. Soder, PHEV home-charging model based on residential activity patterns. IEEE Trans. Power Syst. **28**(3), 2507–2515 (2013)
9. P. Grahn, K. Alvehag, L. Soder, PHEV utilization model considering type-of-trip and recharging flexibility. IEEE Trans. Smart Grid **5**(1), 139–148 (2014)
10. A. Ito, A. Kawashima, T. Suzuki, S. Inagaki, T. Yamaguchi, Z. Zhou, Model predictive charging control of in-vehicle batteries for home energy management based on vehicle state prediction. IEEE Trans. Control Syst. Technol. **26**(1), 51–64 (2018)
11. E.B. Iversen, J.M. Morales, H. Madsen, Optimal charging of an electric vehicle using a Markov decision process. Appl. Energy **123**(15), 1–12 (2014)

12. S. Kamboj, W. Kempton, K.S. Decker, Deploying power grid-integrated electric vehicles as a multi-agent system, in *The 10th International Conference on Autonomous Agents and Multiagent Systems—Volume 1 (AAMAS '11)* (2011), pp. 13–20
13. W. Kempton, J. Tomic, Vehicles-to-grid power implementation: from stabilizing the grid to supporting large-scale renewable energy. J. Power Sources **144**, 280–294 (2005)
14. L. Kessler, K. Bogenberger, Mobility patterns and charging behavior of BMW i3 customers, in *2016 IEEE 19th International Conference on Intelligent Transportation Systems (ITSC)*, Rio de Janeiro (2016), pp. 1994–1999
15. T.-K. Lee, B. Adornato, Z.S. Filipi, Synthesis of real-world driving cycles and their use for estimating PHEV energy consumption and charging opportunities: case study for Midwest/U.S.. IEEE Trans. Veh. Technol. **60**(9), 4153–4163 (2011)
16. W.-H. Lin, J. Zeng, L. Zeng, Experimental study of real-time bus arrival time prediction with GPS data. Transp. Res. Rec. J. Transp. Res. Board **1666**(1), 101–109 (1999). https://doi.org/10.3141/1666-12
17. M. Pantos, Stochastic optimal charging of electric-drive vehicles with renewable energy. Energy **36**(11), 6567–6576 (2011)
18. J. Rice, E. van Zwet, A simple and effective method for predicting travel times on freeways. IEEE Trans. Intell. Transp. Syst. **5**(3), 200–207 (2004)
19. N. Rotering, M. Ilic, Optimal charge control of plug-in hybrid electric vehicles in deregulated electricity markets. IEEE Trans. Power Syst. **26**, 1021–1029 (2011)
20. Y. Sasaki, T. Yamaguchi, A. Kawashima, S. Inagaki, T. Suzuki, Prediction of departure and travel time of individual vehicle based on moving/parked time-passage Markov model and dynamic programming. SICE J. Control Meas. Syst. Integr. **52**(11), 605–613 (2016, in Japanese)
21. S. Shahidinejad, E. Bibeau, S. Filizadeh, Statistical development of a duty cycle for plug-in vehicles in a North American urban setting using fleet information. IEEE Trans. Veh. Technol. **59**, 3710–3719 (2010)
22. E. Sortomme, M.M. Hindi, S.D.J. MacPherson, S.S. Venkata, Coordinated charging of plug-in hybrid electric vehicles to minimize distribution system losses. IEEE Trans. Smart Grid **2**(1), 198–205 (2011)
23. J. Tomic, W. Kempton, Using fleets of electric-drive vehicles for grid support. J. Power Sources **168**, 459–468 (2007)
24. C.-H. Wu, J.-M. Ho, D.T. Lee, Travel-time prediction with support vector regression. IEEE Trans. Intell. Transp. Syst. **5**(4), 276–281 (2004)
25. T. Yamaguchi, S. Inagaki, T. Suzuki, A. Ito, J. Kanamori, Maximum likelihood estimation of departure and travel time of individual vehicle using statistics and dynamic programming, in *IEEE SmartGridComm* (2013), pp. 1352–1358
26. T. Yamaguchi, A. Kawashima, A. Ito, S. Inagaki, T. Suzuki, Real-time prediction for future profile of car travel based on statistical data and greedy algorithm. SICE J. Control Meas. Syst. Integr. **8**(1), 7–14 (2015)

Chapter 3
Design of a Home Energy Management System Integrated with an Electric Vehicle (V2H+HPWH EMS)

Anh Tuan Tran, Akihiko Kawashima, Shinkichi Inagaki, and Tatsuya Suzuki

3.1 Introduction

After the Great East Japan Earthquake in 2011, the amount of electricity generated by nuclear power has declined significantly in Japan. The government also announced a guideline stating that "the dependence on nuclear energy should be reduced as much as possible by introducing power saving/renewable energy and increasing the efficiency of thermal power plants" [10]. Accordingly, it is predicted that renewable energy will become more and more popular in Japan. Among all renewable energy resources, solar power generated through photovoltaics (PV) occupies a large proportion of the green energy consumption in Japan, and the number of residential PV systems is still steadily increasing. Hence, solar power will continue to be a major energy resource in Japan in the future [11]. However, when a large number of PV systems are connected to the power grid, concerns arise on some adverse effects such as voltage or frequency swings due to output fluctuation in PV systems, or surplus electricity and voltage rise issues when the total energy generated by baseload power and renewable energy exceeds the demands. Many solutions have been proposed in order to deal with these issues. Storage batteries have been used in community energy management systems (CEMS) and home

All the figures in this chapter are reprinted with permission of the Institute of Systems, Control and Information Engineers from [16].

A. T. Tran (✉) · A. Kawashima · S. Inagaki · T. Suzuki
Department of Mechanical Systems Engineering, Nagoya University, Nagoya, Aichi, Japan
e-mail: tuan.tran@mae.nagoya-u.ac.jp; inagaki@nagoya-u.jp

T. Suzuki et al. (eds.), *Design and Analysis of Distributed Energy Management Systems*, Power Electronics and Power Systems,
https://doi.org/10.1007/978-3-030-33672-1_3

energy management systems (HEMS) to absorb surplus electricity. In particular, large capacity batteries in electric vehicles (EV) can be used as energy storage units, and their cost and installation space are expected to be reduced [4, 17]. In addition to storage batteries, heat pump water heaters (HPWHs) [21], which have become popular as home appliances in recent years, can be utilized for the same purpose [14]. Furthermore, these two devices can be combined to improve energy absorption efficiency [20] or to deal with both excess power and frequency fluctuation issues [9]. However, in order to promote such systems in the future, it is necessary to maintain a balance between reducing surplus energy to stabilize the power grid system and securing savings and profits for the residents.

The authors in [5, 15, 19] designed a controller for a HPWH system, while taking into account the change in electricity price from daytime to nighttime. However, they considered EV batteries as stationary power storage. In fact, that assumption is unreasonable, since an EV is also a means of transport, which is disconnected from the CEMS/HEMS while traveling. Studies on CEMS/HEMS using in-vehicle storage batteries and HPWH, with the consideration of EV storage battery disconnection, were conducted in [3, 25]. Still, the authors did not design a controller to maximize the economic benefits for residents.

This research aims to develop a model predictive HEMS that optimizes the operation of a HPWH and the charge/discharge profile of an EV to reduce surplus electricity, while taking into account the disconnection of the in-vehicle storage battery, and maximizing the benefits for customers. In this paper, the surplus electricity is defined as the subtraction of the electric power consumption from the total generated power of the household, that is, the electric power sold to the grid. The model predictive HEMS minimizes the electricity expense over the following 24 h with a receding prediction horizon. The advantage of this control method is that it can suppress the effects of modeling and prediction errors, by feeding observation data back at each time step and re-planning. This result has been confirmed in real experiments in [6]. Since we employ many predictions for electric power consumption, hot water demands, generated power, etc., it is difficult to analyze the modeling errors and uncertainties of the HPWH. Therefore, in this research we decided to design the model predictive HEMS. Specifically, the operation plan for the HPWH is incorporated in the control of the charging/discharging process of the in-vehicle battery [24], to reduce the surplus electric power and the electricity bills of customers. The effectiveness of the proposed method is verified using the actual experimental data obtained in Toyota city in Japan [12]. This research describes the optimization problem considered in HEMS and shows some simulation results.

The organization of this chapter is as follows. Section 3.2 presents the mathematical model of a HPWH. Section 3.3 describes the joint optimization problem that simultaneously optimizes the operation plan of the HPWH and the charge/discharge process of the in-vehicle storage battery. Section 3.4 shows simulation results and discussions. Finally, Sect. 3.5 concludes the chapter.

3.2 HPWH Model

This section describes the nonlinear HPWH model proposed by Baba et al. [18]. In addition, the nonlinear model is converted into a piecewise linear one, before applying the optimization problem. However, some modifications in the model, such as the water temperature rise due to heating, are done to meet the research purposes.

3.2.1 *Overview of the HPWH Model*

The description of a HPWH and its mathematical model is shown in the following subsections. HPWH is a device that uses electricity to transfer heat from the surrounding air to the supplied water to heat the water. Due to this feature, the coefficient of performance (COP) of the conversion process from electricity to heat depends on many factors, such as the temperatures of air and water. For example, a HPWH works more efficiently in a warm environment than in a cold one. For a specific HPWH, the rate of heat generation is fixed. Therefore, it is possible to calculate the necessary power consumption to generate an amount of hot water by COP. Also, the volume of generated hot water also depends on the temperature set-point. Since the rate of hot water generation is constant, the quantity produced per time unit decreases when the temperature set-point increases.

Inside the HPWH tank, there are two layers to store hot and cold water. Although the temperature of hot water decreases gradually due to heat loss through the contact wall with the air and cold water, when additional hot water is added to the tank, the water temperature rises to a certain set-point. Furthermore, the mixing of hot and cold water inside the HPWH is not considered. In other words, hot water is withdrawn from the tank and it is mixed with cold water near the faucet to adjust the water temperature for specific purposes.

3.2.2 *Nomenclature*

$\alpha_{T_{\text{set}}}(t) \in \{0, 1\}$	HPWH operation command (0: OFF, 1: ON)
$\alpha_{T_{\text{set}}}^{\text{on}}(t) \in \{0, 1\}$	HPWH start command
$\alpha_{T_{\text{set}}}^{\text{off}}(t) \in \{0, 1\}$	HPWH stop command
Q_{ave}	Average heating capacity [kJ]
$T_{\text{a}}(t)$	Ambient air temperature [°C]
$T_{\text{w}}(t)$	Inlet water temperature
$cal = 4.182$	Specific heat of water [kJ/ℓ K]
$r > 0$	Tank radius [m]
$K_{\text{w}} > 0$	Heat transfer coefficient of water [kJ/K]
$\text{COP}_{\text{ave}} > 0$	Rated coefficient of performance [kJ/kW]

$K_a > 0$	Heat transfer coefficient to the surrounding air [kJ/K m^2]
$P_{HP}(t) \geq 0$	Power consumption of the HPWH [kW]
$T_{set}(t) \in \{65, 90\}$	Temperature set-point [°C]
$L_{use}(t) \geq 0$	Amount of hot water in the tank [ℓ]
$COP(t) \geq 0$	Coefficient of performance [kJ/kW]
$c_1 \sim c_4 \geq 0$	Parameters of COP
$Q_{HPWH}(t) \geq 0$	Amount of heat generated by the HPWH [kJ]
$Q_{st}(t) \geq 0$	Amount of heat stored in the tank [kJ]
$Q_{use}(t) \geq 0$	Amount of heat used by household [kJ]
$Q_a(t) \leq 0$	Heat loss to the air [kJ]
$Q_w(t) \leq 0$	Heat loss to the cold water layer [kJ]
$T_{hw}(t) \geq 0$	Hot water temperature in the tank [°C]
$S_{hw}(t) \geq 0$	Surface area of hot water layer [m^2]
$L_{st}(t) \geq 0$	Amount of hot water in the tank [ℓ]
$L(t) \geq 0$	Amount of generated hot water [ℓ]

Here, $*(t)$ denotes the value at the time t. For instance, $T_a(t)$ denotes the air temperature at the time t. However, $L_{use}(t)$, $Q_{HPWH}(t)$, $Q_{use}(t)$, $Q_a(t)$, $Q_w(t)$, and $L(t)$ represent the amounts of consumed/generated heat or water between the time t and the previous time step. $\alpha_{65}(t)$ and $\alpha_{90}(t)$ indicate the operating conditions of the HPWH between the time t and the previous time step.

3.2.3 Mathematical Model of the HPWH

This subsection describes the mathematical model of the HPWH. The equations are summarized as follows:

$$Q_{HPWH}(t) = Q_{ave}(\alpha_{65}(t) + \alpha_{90}(t)), \tag{3.1}$$

$$P_{HP}(t) = \frac{Q_{HPWH}(t)}{COP(t)}, \tag{3.2}$$

$$COP(t) = (c_1 T_a(t) - c_2 T_w(t) - c_3 T_{set}(t) + c_4)COP_{ave}, \tag{3.3}$$

$$T_{set}(t-1) = 65\alpha_{65}(t) + 90\alpha_{90}(t), \tag{3.4}$$

$$Q_{st}(t) = Q_{st}(t-1) + Q_{HPWH}(t) + Q_a(t) + Q_w(t) - Q_{use}(t), \tag{3.5}$$

$$Q_a(t) = \begin{cases} -(T_{hw}(t-1) - T_a(t-1))S_{hw}(t-1)K_a & \text{if } T_{hw}(t-1) \geq T_a(t-1) \\ 0 & \text{otherwise,} \end{cases} \tag{3.6}$$

$$Q_w(t) = -(T_{hw}(t-1) - T_w(t-1))K_w, \tag{3.7}$$

$$L_{st}(t) = \begin{cases} L_{st}(t-1) + L(t) - L_{use}(t) & \text{if } T_{hw}(t-1) \geq T_w(t-1) \\ 0 & \text{otherwise,} \end{cases} \tag{3.8}$$

$$L(t) = \frac{Q_{HPWH}(t)}{(T_{set}(t-1) - T_w(t-1))cal}, \tag{3.9}$$

$$L_{use}(t) = \frac{Q_{use}(t)}{(T_{hw}(t-1) - T_w(t-1))cal}, \tag{3.10}$$

$$S_{hw}(t) = r^2\pi + 2r\pi\frac{\frac{L_{st}(t)}{1000}}{r^2\pi} = r^2\pi + \frac{L_{st}(t)}{500r}, \tag{3.11}$$

$$T_{hw}(t) = \begin{cases} \frac{(L_{st}(t)-L(t))T_{hw}(t-1)}{L_{st}(t)} + \frac{L(t)T_{set}(t-1)}{L_{st}(t)} + \frac{Q_a(t)+Q_w(t)}{L_{st}(t)cal} & \text{if } L_{st}(t) > 0 \\ T_w(t) & \text{if } L_{st}(t) = 0 \text{ and } \alpha^{on}_{T_{set}}(t) = 0 \\ T_{set}(t) & \text{if } L_{st}(t) = 0 \text{ and } \alpha^{on}_{T_{set}}(t) = 1. \end{cases} \tag{3.12}$$

From (3.1)–(3.12), the amount of heat stored in the HPWH tank $Q_{st}(t)$, heat loss $Q_a(t)$ and $Q_w(t)$, the remaining hot water $L_{st}(t)$, and so forth, are calculated based on the information of the ON/OFF command $\alpha^{on}_{set}(t)/\alpha^{off}_{set}(t)$ of the HPWH, the temperature set-point $T_{set}(t)$, the amount of heat stored in the tank in the previous time step $Q_{st}(t-1)$, the amount of heat used by users $Q_{use}(t)$, the remaining hot water volume in the tank in the previous time step $L_{st}(t-1)$, and the temperature in the previous time step of the hot water $T_{hw}(t-1)$, the inlet water $T_w(t-1)$, and the air $T_a(t-1)$. Furthermore, (3.8) imposes that when the temperature of the hot water is lower than that of the supplied water, the amount of hot water is assumed as zero.

3.2.4 Piecewise Linearization of the HPWH Model

In the mathematical model of the HPWH described in the previous Sects. 3.6, 3.10, and 3.12 contain nonlinear terms. Therefore, when considering the optimization problem with some system constraints, it becomes a mixed-integer nonlinear optimization problem and a nonlinear programming algorithm is needed. In general, it is difficult to obtain an exact solution for this problem, and in many cases, it takes a great deal of time to perform the calculation. Therefore, in this research, the nonlinear equations in the mathematical model of the HPWH are converted into piecewise linear ones. Thus, a mixed-integer linear programming method can be employed. The next section formulates the optimization problem which considers the piecewise linear model of the HPWH and system constraints.

3.3 Model Predictive HEMS Using an In-Vehicle Storage Battery and a HPWH

In this research, the model predictive control method is used to control the charge/discharge process of an in-vehicle storage battery and the operating command ON/OFF of a HPWH. We have conducted studies on the household power consumption, the profile of departure and travel time of the vehicle (PDTT) and its battery power consumption. Based on those data, a prediction algorithm for the household power consumption is proposed using an auto-regressive model [23], and a prediction algorithm for the PDTT of the vehicle is proposed (see Chap. 2). Although there are other studies on photovoltaic power generation [8], household electricity and hot water consumption profiles [18, 22], and weather forecasts, which can be included to predict future demands while designing the HEMS, they are not employed in this research. Instead, we only focus on demonstrating the effectiveness of using both the in-vehicle storage battery and the HPWH, and assume that the future demands are known. When designing the model predictive HEMS, the prediction horizon consists of 48 steps of 30 min, which are equivalent to 24 h. The definitions of variables and parameters used in the optimization problem are shown below.

T	Total time steps
$\widetilde{W}^{+}(k\|t) > 0$	Household power consumption other than the storage battery and the HPWH [kW]
$\widetilde{W}^{-}(k\|t) < 0$	Generated electric power [kW]
$f^{+}(t) > 0$	Purchase price of electricity [Yen/kWh]
$f^{-}(t) > 0$	Sale price of electricity [Yen/kWh]
$\widetilde{\gamma}(k\|t) \in \{0, 1\}$	Vehicle usage index (0: parking at home, 1: traveling)
$\widetilde{B}^{\mathrm{cons}}(k\|t) \geq 0$	Energy consumed by the vehicle while traveling [kWh]
$B^{\mathrm{ref}}(t) \geq 0$	Minimum battery level set by users for a specific desire [kWh]
$B^{\min}(t)$	Minimum battery level constraint [kWh]
$B^{\mathrm{obs}}(t)$	Observed value of the battery level [kWh]
$T_{\mathrm{a}}^{\mathrm{obs}}(t)$	Ambient air temperature [°C]
$T_{\mathrm{w}}^{\mathrm{obs}}(t)$	Inlet water temperature [°C]
$Q_{\mathrm{st}}^{\mathrm{obs}}(t)$	Heat stored in the tank [kJ]
$L_{\mathrm{st}}^{\mathrm{obs}}(t)$	Hot water volume [ℓ]
$T_{\mathrm{hw}}^{\mathrm{obs}}(t)$	Hot water temperature [°C]
$p(k\|t)$	Charging/discharging power of the storage battery [kW]
$W(k\|t)$	Net power consumption [kW]
Δt	Sample time [h]
$S(k\|t) > 0$	Surplus electric power [kW]
$\chi_{\mathrm{acdc}} > 0$	Conversion efficiency of the AC-DC converter
P_{st}	Standby power of the PCS that converts AC in the storage battery to DC for household appliance usage [kW]

$W_{\max} > 0$	Maximum electric power capacity according to the power purchase agreement [kW]
$p^{\text{char}} > 0$	Maximum charging power between the storage battery and the power converter [kW]
$p^{\text{dis}} < 0$	Maximum discharging power between the storage battery and the power converter [kW]
$b(k\|t)$	Remaining energy in the battery [kWh]
$B^{\max} > 0$	Charging capacity of the storage battery [kWh]
$\eta^{\text{char}}, \eta^{\text{dis}}$	Charge/discharge efficiency of the storage battery
$L_{\text{st}}^{\max}$	Hot water tank capacity [ℓ]
Δt_{sec}	Sample time [s]
$T_{\text{hw}}^{\min}$	Minimum hot water temperature set by users for a specific desire [°C]
$Q_{\text{st}}^{\min}$	Minimum amount of heat storage [kJ]
q_{loss}	Heat loss rate
$L_{\text{st}}^{\text{d}}(k\|t)$	Discretized amount of hot water in the tank [ℓ]
$S_{\text{hw}}^{\text{d}}(k\|t)$	Discretized surface area of hot water layer [m^2]

Here, $*(k|t)$ is the conditional prediction, which denotes the predicted value of the variable $*$ at step k in the prediction horizon based on the current data at the time t.

3.3.1 Formulation of the Optimization Problem

The optimization problem proposed in this research is described as follows. The part related to the power consumption of the battery is constructed based on the development in [24].

Given: $\{\widetilde{W}^{+}(k|t),\ \widetilde{W}^{-}(k|t),\ f^{+}(k),\ f^{-}(k), \widetilde{\gamma}(k|t),\ \widetilde{B}^{\text{cons}}(k|t),\ B^{\text{ref}}(k),\ B^{\min}(k),$
$B^{\text{obs}}(t),\ \widetilde{Q}_{\text{use}}(k|t),\ \widetilde{T}_{\text{a}}(k|t),\ \widetilde{T}_{\text{w}}(k|t), T_{\text{a}}^{\text{obs}}(t),\ T_{\text{w}}^{\text{obs}}(t),\ Q_{\text{st}}^{\text{obs}}(t), L_{\text{st}}^{\text{obs}}(t),$
$T_{\text{hw}}^{\text{obs}}(t)\,\}_{k\in[t+1,t+T]},$

find: $\{p(k|t),\ \alpha_{65}(k|t),\ \alpha_{90}(k|t)\,\}_{k\in[t+1,t+T]},$

which minimize:

$$Z = \sum_{\tau=t+1}^{t+T} F(\tau)W(\tau|t)\Delta t + \omega_1 \sum_{\tau=t+1}^{t+T} \left(\alpha_{65}^{\text{on}}(\tau|t) + \alpha_{90}^{\text{on}}(\tau|t)\right)$$
$$+\omega_2 \sum_{\tau=t+1}^{t+T} |p(\tau|t) - p(\tau-1|t)| + \omega_3 \sum_{\tau=t+1}^{t+T} S(\tau|t)\Delta t, \qquad (3.13)$$

where:

$$F(\tau) = \begin{cases} f^{+}(\tau) & \text{if } W(\tau|t) \geq 0 \\ f^{-}(\tau) & \text{otherwise,} \end{cases} \tag{3.14}$$

subject to: $\forall k \in [t+1, t+T]$,

$$W(k|t) = \widetilde{W}^{+}(k|t) + \widetilde{W}^{-}(k|t) + \chi_{\text{acdc}} p(k|t) + P_{\text{st}} + P_{\text{HP}}(k|t), \tag{3.15}$$

$$W(k|t) \leq W^{\max}, \tag{3.16}$$

$$\widetilde{W}^{+}(k|t) + \chi_{\text{acdc}}\, p(k|t) + P_{\text{st}} + P_{\text{HP}}(k|t) \geq 0, \tag{3.17}$$

$$W(k|t) + S(k|t) \geq 0, \tag{3.18}$$

$$P^{\text{dis}} \leq p(k|t) \leq P^{\text{char}}, \tag{3.19}$$

$$B^{\min}(k) \leq b(k|t) \leq B^{\max}, \tag{3.20}$$

$$b(k|t) \geq B^{\text{ref}}(k), \tag{3.21}$$

$$p(k|t)\widetilde{\gamma}(k|t) = 0, \tag{3.22}$$

$$b(t|t) = B^{\text{obs}}(t), \tag{3.23}$$

$$b(k|t) = b(k-1|t) + (1-\widetilde{\gamma}(k|t))\eta(k|t)p(k|t)\Delta t - \widetilde{\gamma}(k|t)\widetilde{B}^{\text{cons}}(k|t), \tag{3.24}$$

$$\eta(k|t) = \begin{cases} \eta^{\text{char}} & \text{if } \quad p(k|t) \geq 0 \\ \eta^{\text{dis}} & \text{otherwise,} \end{cases} \tag{3.25}$$

$$\widetilde{T}_{\text{a}}(t|t) = T_{\text{a}}^{\text{obs}}(t), \tag{3.26}$$

$$\widetilde{T}_{\text{w}}(t|t) = T_{\text{w}}^{\text{obs}}(t), \tag{3.27}$$

$$Q_{\text{st}}(t|t) = Q_{\text{st}}^{\text{obs}}(t), \tag{3.28}$$

$$L_{\text{st}}(t|t) = L_{\text{st}}^{\text{obs}}(t), \tag{3.29}$$

$$T_{\text{hw}}(t|t) = T_{\text{hw}}^{\text{obs}}(t), \tag{3.30}$$

$$0 \leq L_{\text{st}}(k|t) \leq L_{\text{st}}^{\max}, \tag{3.31}$$

$$\widetilde{Q}_{\text{use}}(k-1|t) \leq Q_{\text{st}}(k-1|t), \tag{3.32}$$

$$T_{\text{hw}}(k|t) \geq T_{\text{hw}}^{\min} \qquad \text{if} \quad \widetilde{Q}_{\text{use}}(k|t) > 0, \tag{3.33}$$

$$Q_{\text{st}}(k|t) \geq Q_{\text{st}}^{\min}, \tag{3.34}$$

$$\alpha_{65}(k|t) + \alpha_{90}(k|t) \leq 1, \tag{3.35}$$

$$\alpha_{T_{\text{set}}}(k|t) - \alpha_{T_{\text{set}}}(k-1|t) = \alpha_{T_{\text{set}}}^{\text{on}}(k|t) - \alpha_{T_{\text{set}}}^{\text{off}}(k|t), \tag{3.36}$$

$$\text{COP}(k|t) = (c_1\widetilde{T}_{\text{a}}(k|t) - c_2\widetilde{T}_{\text{w}}(k|t) - c_3T_{\text{set}}(k|t) + c_4)\text{COP}_{\text{ave}}\Delta t_{\text{sec}}, \tag{3.37}$$

$$Q_{\text{HPWH}}(k|t) = Q_{\text{ave}}(\alpha_{65}(k|t) + \alpha_{90}(k|t))\Delta t_{\text{sec}}, \tag{3.38}$$

$$P_{\mathrm{HP}}(k|t) = \frac{Q_{\mathrm{HPWH}}(k|t)}{\mathrm{COP}(k|t)}, \tag{3.39}$$

$$T_{\mathrm{set}}(k|t) = 65\alpha_{65}(k|t) + 90\alpha_{90}(k|t), \tag{3.40}$$

$$L(k|t) = \frac{Q_{\mathrm{HPWH}}(k|t)}{(T_{\mathrm{set}}(k|t) - \widetilde{T}_{\mathrm{w}}(k|t))cal}, \tag{3.41}$$

$$Q_{\mathrm{st}}(k|t) = (1-q_{\mathrm{loss}})Q_{\mathrm{st}}(k-1|t) + Q_{\mathrm{HPWH}}(k-1|t) - \widetilde{Q}_{\mathrm{use}}(k-1|t), \tag{3.42}$$

$$L_{\mathrm{st}}(k|t) = \begin{cases} L_{\mathrm{st}}(k-1|t) + L(k-1|t) - L_{\mathrm{use}}(k-1|t) \\ \qquad \text{if } T_{\mathrm{hw}}(k-1|t) > \widetilde{T}_{\mathrm{w}}(k-1|t) \\ 0 \qquad \text{otherwise,} \end{cases} \tag{3.43}$$

$$L_{\mathrm{use}}(k|t) = \tag{3.44}$$

$$\begin{cases} \frac{\widetilde{Q}_{\mathrm{use}}(k|t)}{(48-\widetilde{T}_{\mathrm{w}}(k|t))cal} - \frac{\widetilde{Q}_{\mathrm{use}}(k|t)(T_{\mathrm{hw}}(k|t)-48)}{(48-\widetilde{T}_{\mathrm{w}}(k|t))^2 cal} & \text{if } T_{\mathrm{hw}}(k|t) \leq 53 \\ \frac{\widetilde{Q}_{\mathrm{use}}(k|t)}{(60-\widetilde{T}_{\mathrm{w}}(k|t))cal} - \frac{\widetilde{Q}_{\mathrm{use}}(k|t)(T_{\mathrm{hw}}(k|t)-60)}{(60-\widetilde{T}_{\mathrm{w}}(k|t))^2 cal} & \text{if } 53 < T_{\mathrm{hw}}(k|t) < 68 \\ \frac{\widetilde{Q}_{\mathrm{use}}(k|t)}{(80-\widetilde{T}_{\mathrm{w}}(k|t))cal} - \frac{\widetilde{Q}_{\mathrm{use}}(k|t)(T_{\mathrm{hw}}(k|t)-80)}{(80-\widetilde{T}_{\mathrm{w}}(k|t))^2 cal} & \text{if } 68 \leq T_{\mathrm{hw}}(k|t), \end{cases} \tag{3.45}$$

$$T_{\mathrm{hw}}(k|t) = \tag{3.46}$$

$$\begin{cases} T_{\mathrm{hw}}(k-1|t) - \frac{L(k-1|t)}{L_{\mathrm{st}}^{\mathrm{d}}(k|t)} T_{\mathrm{hw}}(k-1|t) + \frac{L(k-1|t)}{L_{\mathrm{st}}^{\mathrm{d}}(k|t)} T_{\mathrm{set}}(k-1|t) \\ -\frac{S_{\mathrm{hw}}^{\mathrm{d}}(k-1|t)K_{\mathrm{a}}}{L_{\mathrm{st}}^{\mathrm{d}}(k|t)cal} T_{\mathrm{hw}}(k-1|t) + \frac{S_{\mathrm{hw}}^{\mathrm{d}}(k-1|t)K_{\mathrm{a}}\widetilde{T}_{\mathrm{a}}(k-1|t)}{L_{\mathrm{st}}^{\mathrm{d}}(k|t)cal} \\ -\frac{K_{\mathrm{w}}}{L_{\mathrm{st}}^{\mathrm{d}}(k|t)cal} T_{\mathrm{hw}}(k-1|t) + \frac{K_{\mathrm{w}}\widetilde{T}_{\mathrm{w}}(k-1|t)}{L_{\mathrm{st}}^{\mathrm{d}}(k|t)cal} & \text{if } L_{\mathrm{st}}(t) > 0 \\ \widetilde{T}_{\mathrm{w}}(k|t) & \text{if } L_{\mathrm{st}}(t) = 0 \text{ and } \alpha_{T_{\mathrm{set}}}^{\mathrm{on}}(t) = 0 \\ T_{\mathrm{set}}(k|t) & \text{if } L_{\mathrm{st}}(t) = 0 \text{ and } \alpha_{T_{\mathrm{set}}}^{\mathrm{on}}(t) = 1, \end{cases} \tag{3.47}$$

$$\begin{cases} L_{\mathrm{st}}^{\mathrm{d}}(k|t) = 0, \quad S_{\mathrm{hw}}^{\mathrm{d}}(k|t) = 0 & \text{if } L_{\mathrm{st}}(k|t) = 0 \\ L_{\mathrm{st}}^{\mathrm{d}}(k|t) = 50, \quad S_{\mathrm{hw}}^{\mathrm{d}}(k|t) = 0.59 & \text{if } 0 < L_{\mathrm{st}}(k|t) \leq 100 \\ L_{\mathrm{st}}^{\mathrm{d}}(k|t) = 200, \quad S_{\mathrm{hw}}^{\mathrm{d}}(k|t) = 1.75 & \text{if } 100 < L_{\mathrm{st}}(k|t) \leq 300 \\ L_{\mathrm{st}}^{\mathrm{d}}(k|t) = 350, \quad S_{\mathrm{hw}}^{\mathrm{d}}(k|t) = 2.9 & \text{if } 300 < L_{\mathrm{st}}(k|t) \leq 370. \end{cases} \tag{3.48}$$

The parameters in the equations above are summarized in Table 3.1. Furthermore, the values of χ_{ACDC}, P_{st}, W_{max}, P^{char}, P^{dis}, B^{max}, B^{min}, η^{char}, η^{dis} are derived from an actual HEMS experiment system, the values of $c_1, \ldots, c_4$, K_a, K_w are given in [18], and the value of q_{loss} is cited in [5]. As for the values of $L_{\mathrm{st}}^{\mathrm{max}}$, $\mathrm{COP}_{\mathrm{ave}}$, and Q_{ave}, they can be found in the HPHW manual.

Table 3.1 Parameter settings

T	48	Δt	0.5
ω_1	1	ω_2	1×10^{-4}
ω_3	200	χ_{ACDC}	1.03
P_{st}	1.55×10^{-1}	W_{max}	6
p^{char}	3	p^{dis}	−3
B^{max}	24	B^{min}	7.2
η^{char}	0.96	η^{dis}	1.04
L_{st}^{max}	370	T_{hw}^{min}	45
Q_{st}^{min}	8000	c_1	1.6×10^{-2}
c_2	1.16×10^{-2}	c_3	8.3×10^{-3}
c_4	1.47	COP_{ave}	4
Δt_{sec}	1800	Q_{ave}	4.5
q_{loss}	5×10^{-3}	K_a	1×10^{-3}
K_w	5.27×10^{-4}	–	–

3.3.2 Cost Function

The first term in (3.13) represents the electricity charge in the following 24 h. In (3.14), the purchase and sale electricity price are considered. The second term in (3.13) indicates the number of activation times of the HPWH in the following 24 h. In other words, it shows how many times the HPWH system switches from OFF to ON in the prediction horizon. By taking this term into account, it can prevent the system from frequently switching between ON and OFF modes and thus reduce the energy loss during the start-up period. The coefficient ω_1 represents the 10% energy loss during the start-up according to [5], which contributes to the total household energy consumption cost (approximately 1 yen per start-up). The third term is to prevent the battery from frequently switching between charging and discharging states; therefore, it helps to improve the battery lifespan. The coefficient ω_2 is set as a small number, so that the third term only has a significant effect if the state of charge of the battery repeatedly changes. The fourth term is a penalty for surplus electric power, which is to reduce the amount of electricity sold to the grid. Since the purpose of this research is to reduce the surplus electric power before minimizing the household electricity bills, the coefficient ω_3 takes a large value so that the third term plays a major role in the cost function compared with the other ones. However, when modifying the value of ω_3, the balance between excess power suppression and household electricity cost minimization is also changed to a great extent. Therefore, an investigation into the effect of ω_3, together with an examination of remuneration to customers for excess power suppression, should be further studied in the future.

3.3.3 Constraints on Household Electric Power Consumption

Equation (3.15) calculates the net power consumption. It indicates the purchased amount of electric power if the value is positive, or the amount sold to the grid if the value is negative. Equation (3.16) represents the upper constraints on the power consumption. Equation (3.17) expresses that the storage battery cannot discharge more power than the amount consumed by home appliances, which implies that the electricity from the storage battery cannot be sold to the grid. Equation (3.18) calculates the surplus electric power $S(k|t)$. Together with the fourth term of the cost function, $S(k|t)$ is equal to $-W(k|t)$ only when $W(k|t) < 0$.

3.3.4 Constraints on In-Vehicle Storage Batteries

Equation (3.19) expresses the charge/discharge amount of the storage battery. Equation (3.20) indicates the upper and lower bounds of the storage battery capacity. Equation (3.21) is a constraint on the minimum battery level at departure to meet the customer's needs. Equation (3.22) is a constraint related to the charge/discharge amount of the in-vehicle battery. It can be clearly seen that the value equals zero when the car is being used. Equations (3.23) and (3.24) represent the observed values of the current and predicted battery levels, respectively. Furthermore, the charge/discharge efficiency of the storage battery at each time step is calculated in (3.25).

3.3.5 Constraints on the HPWH

Equations (3.26)–(3.33) match the predicted values of the temperature, thermal energy, and hot water volume at the current time with the observed data. Equation (3.31) is the upper and lower bounds of the hot water level in the tank. Equation (3.32) is a constraint on the amount of heat energy used, which means that the consumed heat should be smaller than the one stored in the hot water tank. Equation (3.33) is a constraint on the temperature of hot water in the tank. It indicates that when the hot water is used, the temperature of the remaining hot water in the tank is planned to be equal to or higher than the minimum value $T_{\mathrm{hw}}^{\mathrm{min}}$. In general, the hot water temperature is usually above 43 °C. However, in this research, the value of $T_{\mathrm{hw}}^{\mathrm{min}}$ is set at 45 °C to compensate for modeling errors and uncertainties. Equation (3.34) is a constraint on the minimum amount of heat stored in the tank at every time step. In this research, the value of $Q_{\mathrm{st}}^{\mathrm{min}}$ is set to be 8000 kJ. Equation (3.35) is to prevent the HPWH from choosing the set-points 65 and 90 °C simultaneously. Equation (3.36) determines the values of α^{on} and α^{off}. In the step that α switches from 0 to 1, α^{on} is 1 meanwhile α^{off} is 1 in the step that α switches from 1 to 0. Equations (3.37)–(3.48) represent the HPWH model.

Equations (3.42), (3.45), and (3.47) and (3.48) are the approximations of (3.6), (3.10), and (3.12) described in the previous section, respectively. According to [5], (3.6) and (3.7) are simplified, so that the heat loss only depends on the amount of stored heat in the previous step. Regarding Eq. (3.10), the model for hot water temperature in the tank, T_{hw}, is linearized around three operating points using the Taylor expansion method, and those three models are switched during the operation. Finally, (3.12) is expressed in a piecewise, linear fashion by discretizing L_{st} and S_{hw} and switching the values. In order to investigate the influence of approximation errors on the results due to the piecewise linearization, the values of Q_{st}, L_{st}, and T_{hw} calculated using the original nonlinear model and the piecewise linear one, under the same operation pattern for the HPWH in 1 week, are compared. Table 3.2 shows the average relative error between the nonlinear model and the piecewise linear one. As shown in the table, the approximation error for each variable is about 10% at most. In addition, Fig. 3.1 illustrates the transition of Q_{st} over 1 week. It can be seen that the error gradually increases, but does not diverge. Since the model predictive control method used in this research can correct the modeling error every

Table 3.2 Relative error between the original nonlinear model and the approximated piecewise linear one

Q_{st}	L_{st}	T_{hw}
9.81%	11.9%	1.78%

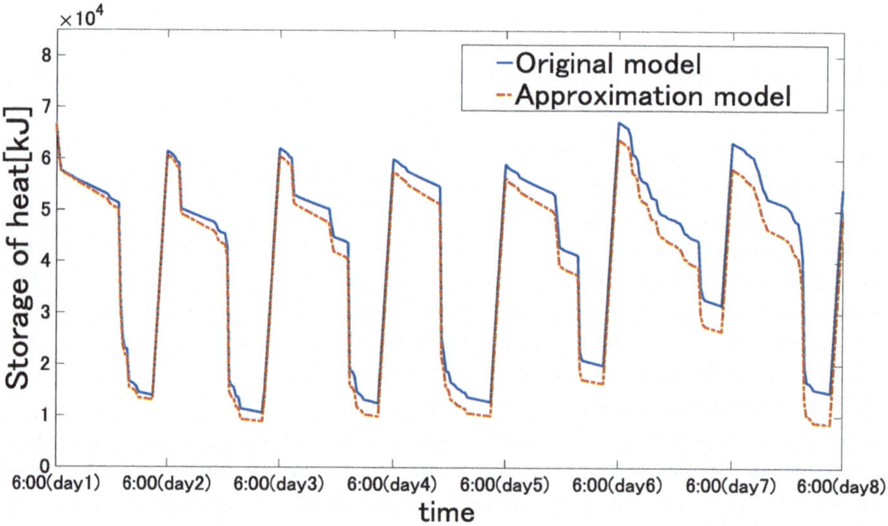

Fig. 3.1 Difference in Q_{st} obtained by the original nonlinear model and the approximated piecewise linear one [16]

30 min (one step), the impact of the approximation error on the optimization result can be neglected. Moreover, some equations, such as (3.14), can be converted into MLDS using logical variables. For more details, please refer to [1].

3.4 Numerical Simulations

3.4.1 *Simulation Settings*

In this section, numerical simulations are conducted in order to confirm the effectiveness of the proposed HEMS on reducing the household electricity charges and surplus electric power, while considering the detachment of the in-vehicle storage battery from the HEMS. Moreover, we propose a sequential optimization approach in addition to the joint optimization described above, to reduce the computation time. In the sequential optimization approach, the operation plan for the HPWH is optimized without considering the vehicle batteries. Next, based on the obtained operation plan, the prediction of the household energy consumption and the generated electric power, we calculate the optimal charge/discharge plan for the in-vehicle battery. This method is compared with the conventional one, in which the HPWH operation is assumed to be carried out at night, which is reasonable at the present, and the optimization is done only for the in-vehicle storage battery. As for the operation of HPWH considered in the conventional method, the operation start-up time is calculated backward, so that the storage tank is filled with hot water at 6 a.m. Table 3.3 shows the specifications of the computer and the optimization solver used in the simulations.

In this research, the simulations are conducted over a 1-week period. The household can control the charge/discharge process of the EV, the ON/OFF operation of the HPWH, and the generated electricity from the PV system. We use the actual data of power consumption, thermal energy usage, and generated electricity for a four-person family in Toyota City, Japan, in May [12]. The real air temperature in the same day is also obtained from the Japan Meteorological Agency database [7]. As for the electricity price, we used the purchase price of the Chubu Electric Power E Life Plan [2] (see Fig. 3.2) and fixed sale prices, which is anticipated in [13]. There are three car patterns which are used for different purposes: Car A is mainly used for commuting, Car B is used for going to the part-time job location, and Car C is used for shopping purposes. The departure and return time profile, the power

Table 3.3 Computer specifications

OS	Windows 7 professional
CPU	Intel(R) Core(TM) i7-2600S CPU@2.8 GHz
Memory	8 GB
Solver	Gurobi 6.5

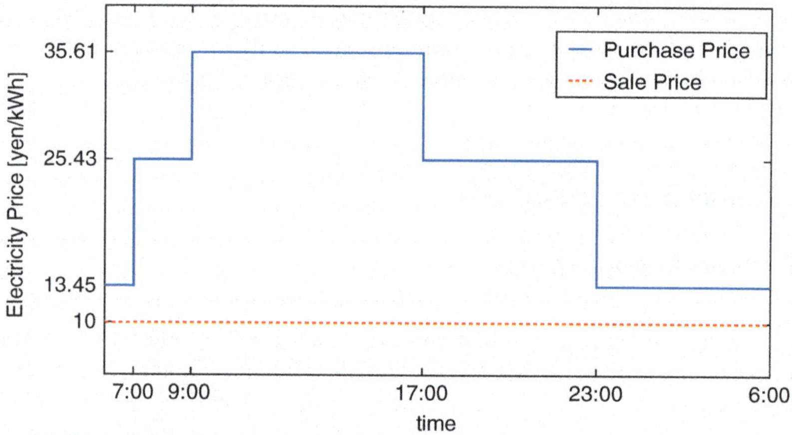

Fig. 3.2 Electricity price data [16]

Table 3.4 Patterns of car usage (Car A)

Departure time	Return time	Power consumption	Lower boundary
8:00	20:00	6 kWh	19.2 kWh

Table 3.5 Patterns of car usage (Car B)

Departure time	Return time	Power consumption	Lower boundary
8:00	12:00	1 kWh	12 kWh
15:00	18:00	1 kWh	12 kWh

Table 3.6 Patterns of car usage (Car C)

Departure time	Return time	Power consumption	Lower boundary
7:30	8:30	1 kWh	12 kWh
13:30	15:00	1 kWh	12 kWh
17:00	18:00	1 kWh	12 kWh

consumption, and the desired, minimum battery level set by users, are presented in Tables 3.4, 3.5, and 3.6.

3.4.2 Simulation Results

The average computation time for each step, the accumulated surplus electric power over 1 week, and the electricity bills for 1 week are shown in Tables 3.7, 3.8, 3.9, respectively. The numbers are rounded up to improve readability. The simulation result on the third day is shown in Figs. 3.3, 3.4, and 3.5 as an example. In these figures, the vertical axes represent the amount of electric power, the remaining

Table 3.7 Computation time [s/step]

	Conventional	Sequential	Joint
Car A	1	5	6
Car B	1	5	3
Car C	1	5	74

Table 3.8 Surplus power [kWh/week]

	Conventional	Sequential	Joint
Car A	37.64	19.9	20.12
Car B	24.02	10.72	7.89
Car C	11.96	10.42	4.69

Table 3.9 Electricity charge [Yen/week]

	Conventional	Sequential	Joint
Car A	2779	2835	2809
Car B	1932	1995	1990
Car C	1857	1935	1814

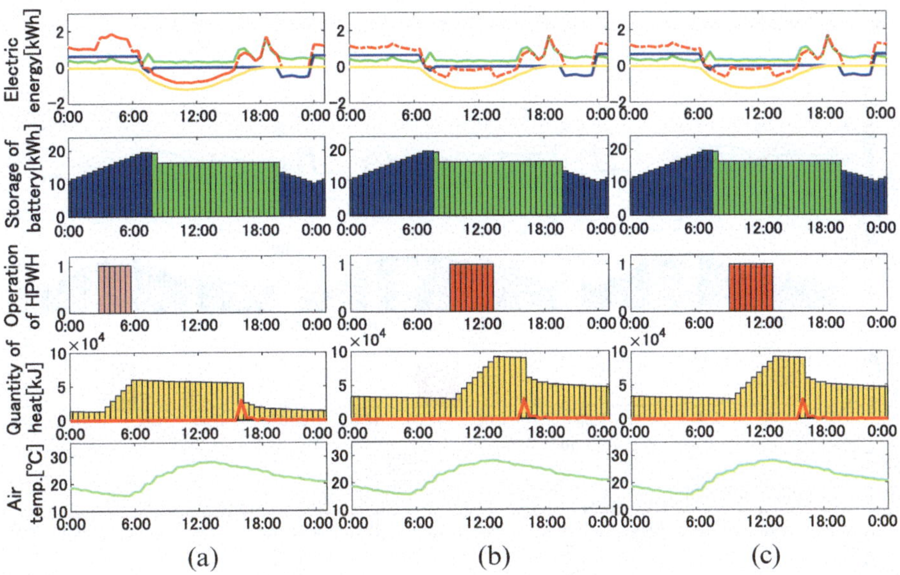

Fig. 3.3 Simulation results (Car A) [16]. (**a**) Conventional. (**b**) Sequential. (**c**) Joint

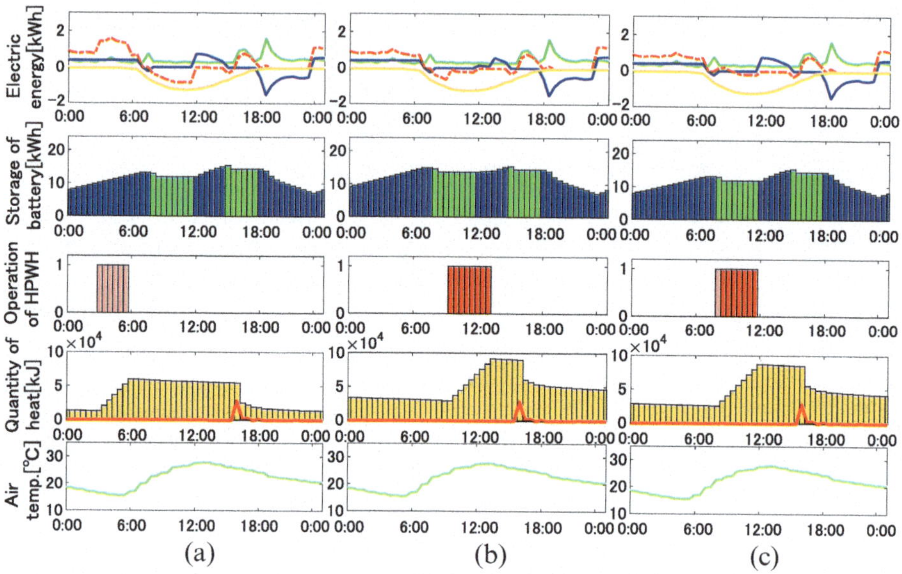

Fig. 3.4 Simulation results (Car B) [16]. (**a**) Conventional. (**b**) Sequential. (**c**) Joint

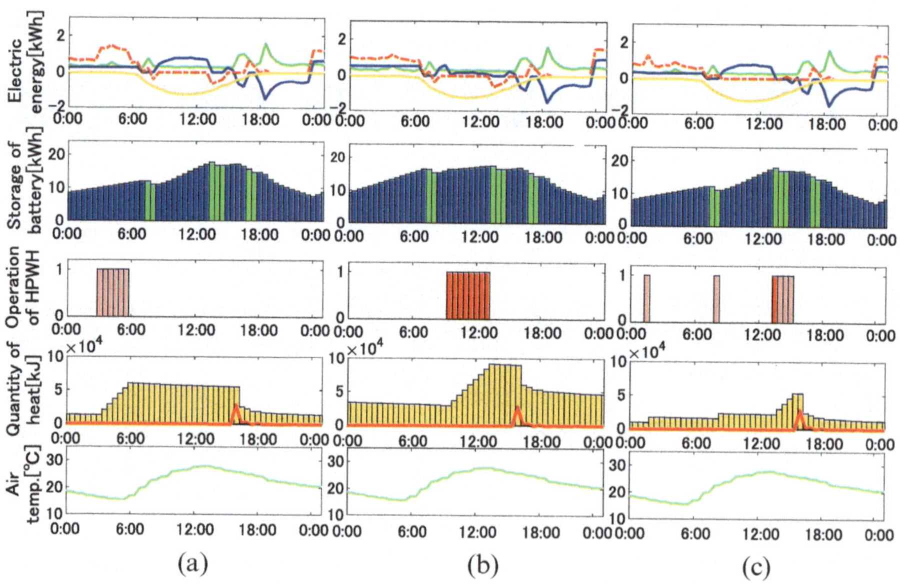

Fig. 3.5 Simulation results (Car C) [16]. (**a**) Conventional. (**b**) Sequential. (**c**) Joint

battery energy, the operation state of the HPWH, the amount of thermal energy, and the ambient air temperature, while the horizontal axes illustrate the time. As for the charge/discharge amount of electric power shown in the first stages of Figs. 3.3, 3.4, and 3.5, positive value means that the battery is being charged and negative one implies the discharge process. The results of three optimization approaches are provided, which are the conventional optimization, sequential optimization, and joint optimization ones.

3.4.3 Discussion

3.4.3.1 Computation Time

As can be seen in Table 3.7, the computation time for both joint optimization and sequential optimization approaches is smaller than the sample time, which is 30 min. Therefore, it confirms that the solutions to the optimization problem can be obtained in a realistic time frame, by linearizing the HPWH model. In addition, the sequential optimization approach takes less time than the joint optimization one. Especially, the calculation time difference between two approaches in the optimization of Car C is more than 1 min. This result shows that when the degree of freedom of the system increases, or when the travel patterns of the car are complicated, the joint optimization method takes more time to analyze solutions than the sequential optimization one. In some cases, such as with Car B, the calculation time of the joint optimization approach is shorter. However, it does not mean that the sequential optimization one is inferior. We can only conclude that the calculation time of the sequential optimization method is more stable and less affected by the system's complexity than that of the joint optimization one.

3.4.3.2 Surplus Electric Power

By looking at the results for Car A in Table 3.8, it can be confirmed that the surplus electric power is suppressed for both sequential optimization and joint optimization methods compared with the conventional one, due to the fact that the HPWH can be operated during daytime to absorb more excess energy. It can be also verified that the HPWH is operated at the temperature set-point of 90 °C, to consume more surplus electric power during daytime, while in the conventional optimization approach, the HPWH is operated at the set-point of 65 °C. Moreover, there is almost no difference in the results of the sequential optimization method and the joint optimization one in Fig. 3.3 since the in-vehicle storage batteries are mostly unavailable during daytime and cannot be used to absorb excess energy.

In the case of Car B, similar results can be obtained. The sequential optimization and the joint optimization approaches can reduce more excess energy than the conventional one. Furthermore, the difference in the results of these two approaches can be observed. As shown in Fig. 3.4, the joint optimization approach activates the HPWH earlier, to absorb surplus electric power during the period when the car is running and the in-vehicle storage battery is disconnected from the HEMS. Furthermore, it can be observed that when the power consumption of the HPWH is greater than the amount of surplus electric power, the operation plan is changed to the set-point of 65 °C, to avoid purchasing unnecessary electric power.

Concerning the operation of Car C, a similar trend on surplus electric power to the case of Car B can be observed. However, the difference between the sequential optimization method and the joint optimization one is more evident. Figure 3.5 shows that in the joint optimization approach, the HPWH is active while the vehicle is traveling, to reduce the excess energy. Furthermore, when the car is parked, the surplus electric power and the electricity of the in-vehicle storage battery are used to heat the water according to customers' demands instead of using the HPWH. Therefore, the joint optimization approach can significantly reduce the surplus electric power, while still satisfying the demands of customers.

3.4.3.3 Electricity Charges

In the sequential optimization and joint optimization approaches, the HPWH is operated during the daytime to absorb surplus electric power, while in the conventional one HPWH is operated at night. Therefore, the electricity charges in the conventional approach are cheaper since the electricity price during the daytime is nearly 3 times higher than that during nighttime. However, since the operation plan can be optimized so as to not affect the electricity charges as much as possible, the difference in electricity bills can be limited. Moreover, during daytime when the temperature is high, the COP has a better value than that during the nighttime, and thus, the HPWH can operate more effectively. It can be considered as a factor for reducing electricity bills.

3.5 Conclusion

In this research, we developed a model predictive HEMS that optimizes the operation of the in-vehicle storage battery and the HPWH, to reduce surplus electric power while taking into account the disconnection of in-vehicle storage batteries from the HEMS and maximizing the benefits for customers. We formulated a joint optimization problem for the charge/discharge plan of in-vehicle storage batteries and the operation of the HPWH. In addition, the nonlinear HPWH model is converted into a piecewise linear one in order to obtain the solution for the optimization problem in a more rapid, and realistic time frame. The numerical simulations using the actual demand data for a family from Toyota City in Japan, demonstrated that the complementary operation of HPWH and in-vehicle storage batteries can reduce the surplus electric power and electricity charges, while considering the disconnection of in-vehicle storage batteries from the HEMS. Notably, if the vehicle is connected to the HEMS for a long time during the daytime, the benefits of customers can be increased.

As for future tasks, it is necessary to further examine the proposed method while taking into account additional prediction models that are not considered in this research, such as household electricity and hot water consumption profiles or weather forecasts. Also, the balance between surplus electric power suppression and benefits for customers should be further investigated.

Acknowledgements This research was supported by JST CREST Grant Number JPMJCR15K3. The authors would like to thank DENSO Corporation for valuable discussions and experimental data. The authors also would like to thank Professor Shunpei Baba of the Graduate School of Frontier Sciences at the University of Tokyo for his valuable comments.

References

1. K. Asano, K. Tsuda, A. Bemporad, M. Morari, Predictive control of hybrid system and its application to process control. Trans. Inst. Syst. Control Inform. Eng. **46**(3), 110–119 (2002, in Japanese)
2. Chubu Electric Power E Life Plan (three-time-zone electricity plan)—Basic menu (electricity contract) (in Japanese). http://www.chuden.co.jp/home/home_menu/home_basic/hba_elife/index.html
3. Y. Hanai, K. Yoshimura, J. Matsuki, Y. Hayasi, Load management using heat pump water heater and electric vehicle battery charger in distribution system with PV. J. Int. Council Electr. Eng. **1**(2), 2007–2013 (2011)
4. K. Hirata, T. Kamiji, A distributed EV/PHEV storage management strategy for power flow balancing. Trans. Inst. Syst. Control Inform. Eng. **28**(11), 427–434 (2015) (in Japanese)
5. T. Ikegami,Y. Iwafune, K. Ogimoto, Development of the optimum operation scheduling model of domestic electric appliances for the supply-demand adjustment in a power system. IEEJ Trans. Power Energy **130**(10), 877–887 (2010, in Japanese)
6. A. Ito, A. Kawashima, T. Suzuki, S. Inagaki, T. Yamaguchi, Z. Zhou, Model predictive charging control of in-vehicle batteries for home energy management based on vehicle state prediction. IEEE Trans. Control Syst. Technol. **26**(1), 51–64 (2018)
7. Japan Meteorological Agency—weather data history (in Japanese). http://www.data.jma.go.jp/obd/stats/etrn/index.php?sess=6ef525a9cdef28cea634ce58ca736e68
8. M. Kudo, A. Takeuchi, Y. Nozaki, H. Endo, J. Sumita, Forecasting electric power generation of photovoltaic power system for energy network. IEEJ Trans. Power Energy **127**(7), 847–853 (2007, in Japanese)
9. T. Masuta, A. Yokoyama, Supplementary load frequency control by use of a number of both electric vehicles and heat pump water heaters. IEEE Trans. Smart Grid **3**(3), 1253–1262 (2012)
10. Ministry of Economy, Trade and Industry, Energy Basic Plan (2014) (in Japanese). http://www.meti.go.jp/press/2014/04/20140411001/20140411001-1.pdf
11. Ministry of Economy, Trade and Industry—Agency for Natural Resources and Energy (2016) Annual report on energy in FY2007 (in Japanese). http://www.enecho.meti.go.jp/about/whitepaper/2016pdf/whitepaper2016pdf_2_1.pdf
12. Ministry of Economy, Trade and Industry, Next generation energy and social systems demonstration project—final report (Toyota city), in *The 18th Conference on Next Generation Energy and Social Systems* (2016, in Japanese). http://www.meti.go.jp/committee/summary/0004633/pdf/018_06_02.pdf

13. Ministry of Economy, Trade and Industry Procurement Price Calculation Committee (2016) Handouts on January 19, 2016 (in Japanese). http://www.meti.go.jp/committee/chotatsu_kakaku/pdf/020_01_00.pdf
14. Y. Miyamoto, Y. Hayashi, Prevention of output suppression through heat pump water heaters for high-penetration residential PV systems for long-term operation. IEEJ Trans Power Energy **135**(7), 426–436 (2015, in Japanese). https://doi.org/10.1541/ieejpes.135.423
15. Y. Miyamoto, Y. Hayashi, Effectiveness of improvement of benefit of the electric power selling charges and electricity charges using battery storage system and heat pump water heater for high penetration PV. IEEJ Trans. Power Energy **136**(3), 245–258 (2016, in Japanese)
16. I. Nawata, Y. Sasaki, A. Kawashima, S. Inagaki, T. Suzuki, Cooperative operation of in-vehicle battery and HPWH in model predictive HEMS. Trans. Inst. Syst. Control Inform. Eng. **30**(4), 143–152 (2017, in Japanese)
17. T. Nishioeda, M. Suzuki, N. Morino, T. Oka, Y. Takebayashi, Load leveling by using electric vehicle batteries to store surplus energy generated by photovoltaic systems on single-family houses in Japan. J. Environ. Eng. **80**(708), 149–158 (2015, in Japanese)
18. K. Okutani, J. Baba, Y. Ota, Fundamental study on demand and grid side impact by use of a heat pump water heater for household as a controllable load. IEEJ Trans. Power Energy **136**(1), 72–78 (2016, in Japanese)
19. E. Omine, H. Hatta, M. Asari, T. Ueno, H. Kobayashi, Operation method of autonomous demand area power system to utilize surplus power of PV systems by using heat pump hot water heaters and electric energy storage systems. IEEJ Trans. Power and Energy **133**(7), 631–641 (2012, in Japanese)
20. Y. Saito, Y. Tomita, K. Imai, M. Tsurugai, S. Tamura, Trial of carbon-free micro-grid project in Japan, in *IEEE International Conference on Smart Grid Engineering (SGE)* (2012), pp. 1–5
21. The Japan Refrigeration and Air Conditioning Industry Association, Changes of the domestic shipment number of the heat pump water heater (2015, in Japanese). http://jraia.or.jp/statistic/s_heatpump.html
22. J. Widen, M. Lundh, I. Vassileva, E. Dahlquist, K. Ellegard, E. Wackelgard, Constructing load profiles for household electricity and hot water from time-use data-modelling approach and validation. Energy Build. **41**(7), 753–768 (2009)
23. T. Yamaguchi, A. Kawashima, A. Ito, S. Inagaki, T. Suzuki, Real-time prediction for future profile of car travel based on statistical data and greedy algorithm. SICE J. Control Measur. Syst. Integr. **8**(1), 7–14 (2013)
24. T. Yamaguchi, M. Sumiya, S. Inagaki, T. Suzuki, A. Ito, M. Fujita, J. Kanamori, Model predictive control of car storage battery in HEMS considered car traveling, in *Proceedings of SICE Annual Conference* (2013), pp. 1352–1358
25. Q. Zhang, T. Tezuka, K. Ishihara, B. Mclellan, Integration of PV power into future low-carbon smart electricity systems with EV and HP in Kansai Area, Japan. Renew. Energy **44**, 99–108 (2012)

Chapter 4
Range Extension Autonomous Driving for Electric Vehicles Based on Optimization of Velocity Profile Considering Traffic Signal Information

Hiroshi Fujimoto, Hideki Yoshida, and Yoshi Ri

4.1 Introduction

Electric vehicles (EVs) have been attracting widespread attention as one of the solutions to global environmental and energy problems. Moreover, EVs are superior to internal combustion engine vehicles (ICEVs) in terms of motion control [6]. The mileage per charge of EVs, however, remains lower than the mileage of ICEVs. To address this issue, various methods were proposed [2, 5, 10–12]. For example, in [5], motor efficiency of EVs is improved by realizing current control for individual windings. In [2], the authors proposed an optimization algorithm to distribute front and rear driving/braking forces by considering the slip ratio and motor loss. However, the above-mentioned studies focused on vehicles driven by human drivers.

Owing to the advances in intelligent transport systems (ITS), infrastructure-to-vehicle (I2V) communications can provide a fast and cheap advanced driving assistance system (ADAS) [9]. In the near future, vehicle velocity will be determined by autonomous driving systems based on ITS information and traffic environment. Various research groups have proposed the optimization of the vehicle velocity profile to reduce the total energy consumption [8, 14]. However, most of the research did not deal with the short mileage per charge characteristic of EVs. In [14], the energy consumption was reduced by optimizing the velocity profile and considering the traffic signal information based on the assumption that ITS provides

All the figures in this chapter are reprinted with permission of the Institute of Electrical and Electronics Engineers from [3].

H. Fujimoto (✉) · H. Yoshida · Y. Ri
Graduate School of Frontier Sciences, University of Tokyo, Tokyo, Japan
e-mail: fujimoto@k.u-tokyo.ac.jp

T. Suzuki et al. (eds.), *Design and Analysis of Distributed Energy Management Systems*, Power Electronics and Power Systems,
https://doi.org/10.1007/978-3-030-33672-1_4

the traffic signal information in advance. However, these studies did not consider iron loss and slip loss occurring in EVs.

Our research group has previously proposed the range extension autonomous driving (READ) system using only motion control techniques [3, 4, 7], which does not require a change in motor type and vehicle structure. READ can reduce energy consumption by optimizing the velocity profile of the vehicle. However, conventional READ did not consider important factors, such as traffic signal information. READ based on additional traffic signal information can be expected to reduce energy consumption significantly. In this study, the traffic signal information is assumed to be available from optical beacons, and the velocity profile can be designed by solving a nonlinear optimal control problem. The effectiveness of the proposed method is verified by numerical simulations and experiments.

4.2 Experimental Vehicle and Its Mathematical Model

4.2.1 Experimental Vehicle

In this research, an original electric vehicle "FPEV2-Kanon", manufactured by our research group, is used to test the modified READ system. The vehicle and its specifications are illustrated in Fig. 4.1 and Table 4.1, respectively. This vehicle has four outer-rotor type in-wheel motors, which are of the direct drive type. Therefore, the reaction forces from the road are directly transferred to the motor without the backlash effect of the reduction gear. Table 4.2 and Fig. 4.2 list the specifications of the motors and the efficiency maps of the front and rear in-wheel motors, respectively. Lithium-ion batteries are used as power source for the motors. The

Fig. 4.1 FPEV2-Kanon [3]

Table 4.1 Vehicle specifications

Meaning	Symbol	Value
Vehicle mass	M	854 kg
Wheel base	l	1.72 m
Distance from CG to front axle	l_f	l_f:1.01 m
Distance from CG to rear axle	l_r	l_r:0.702 m
Front wheel inertia	J_{ω_f}	1.24 kg m^2
Rear wheel inertia	J_{ω_r}	1.26 kg m^2
Wheel radius	r	0.302 m

Table 4.2 Specifications of in-wheel motors

	Front	Rear
Manufacturer	TOYO DENKI SEIZO K.K.	
Type	Direct drive system	
	Outer rotor type	
Rated torque	110 Nm	127 Nm
Maximum torque	500 Nm	530 Nm
Rated power	6.00 kW	6.00 kW
Maximum power	20.0 kW	25.0 kW
Rated speed	382 rpm	450 rpm
Maximum speed	1110 rpm	1200 rpm

output voltage of the main battery is 160 V. The voltage is boosted to 320 V by a converter. In this study, the converter loss is assumed to be negligible and can be ignored.

4.2.2 Vehicle Model

In this section, a four-wheel drive vehicle model is described. Based on the model in Fig. 4.3a, the wheel dynamics are expressed as (4.1). From Fig. 4.3b, the vehicle dynamics are expressed as (4.2)–(4.4)

$$J_{\omega_j}\dot{\omega}_j = T_j - rF_j, \tag{4.1}$$

$$M\dot{V} = F_{\mathrm{all}} - \mathrm{sgn}(V)F_{\mathrm{DR}}(V), \tag{4.2}$$

$$F_j = \frac{1}{4}F_{\mathrm{all}}, \tag{4.3}$$

$$F_{\mathrm{DR}}(V) = \mu_0 Mg + b|V| + \frac{1}{2}\rho C_{\mathrm{d}} A V^2, \tag{4.4}$$

where ω_j is the angular velocity of the wheel, V is the vehicle velocity, T_j is the motor torque, F_j is the driving force of each wheel, F_{all} is the total driving force, M is the vehicle mass, r is the radius of the wheel, J_{ω_j} is the inertia experienced by

Fig. 4.2 Efficiency maps of front and rear motors [3]. (**a**) Front motor. (**b**) Rear motor

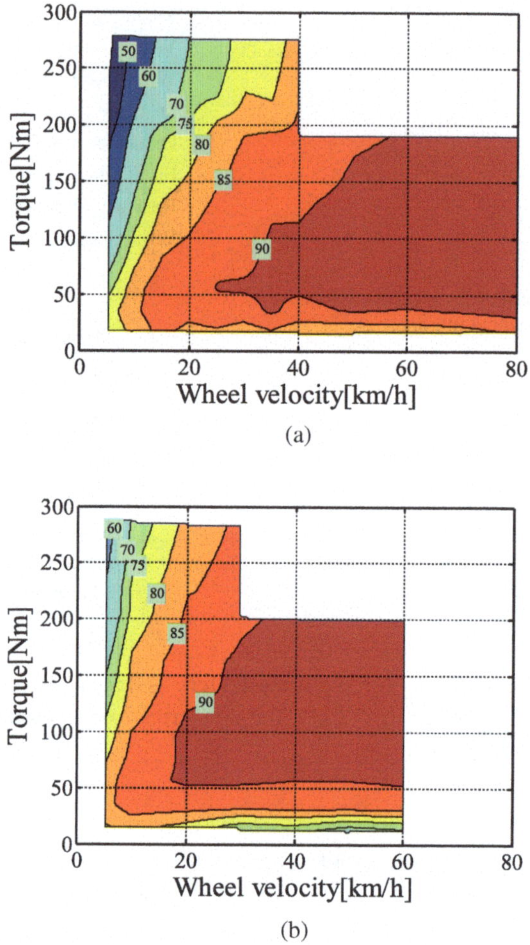

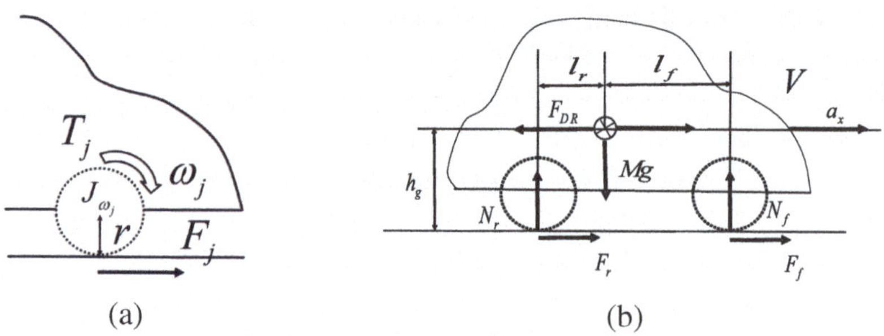

Fig. 4.3 Vehicle model. (**a**) Rotational motion of a wheel. (**b**) Load transfer model

the wheel, F_{DR} is the driving resistance, μ_0 is the rolling friction coefficient, b is the rolling resistance coefficient, ρ is the density of the air surrounding the vehicle, C_{d} is the drag coefficient, and A is the frontal projected area. The subscript j represents f or r, where f indicates "front" and r indicates "rear."

The slip ratio λ_j is defined as

$$\lambda_j = \frac{V_{\omega_j} - V}{\max(V_{\omega_j}, V, \epsilon)}, \tag{4.5}$$

where $V_{\omega_j} = r\omega_j$ is the wheel speed and ϵ is a small constant to avoid zero division. The slip ratio λ_j is known to be related with the friction coefficient μ_j [13]. In the region $|\lambda_j| \ll 1$, μ_j is nearly proportional to λ_j. Therefore, for cases where longitudinal acceleration is present,

$$F_j = \mu_j N_j \approx D_s^{'} N_j \lambda_j, \tag{4.6}$$

where N_f and N_r are the front and rear normal forces, respectively, and $D_s^{'}$ is the normalized driving stiffness.

The normal forces acting on each wheel during the longitudinal acceleration phase are calculated as follows:

$$N_f(\dot{V}) = \frac{1}{2}\left[\frac{l_r}{l}Mg - \frac{h_g}{l}M\dot{V}\right], \tag{4.7}$$

$$N_r(\dot{V}) = \frac{1}{2}\left[\frac{l_f}{l}Mg + \frac{h_g}{l}M\dot{V}\right], \tag{4.8}$$

where l_f and l_r are the distances from the center of gravity to the front and rear axles, respectively, l is the wheelbase, and h_g is the height of the center of gravity from the base of the vehicle.

4.2.3 *Power Flow Model [4]*

The inverter input power P_{in}, which considers the slip ratio and motor loss, is expressed as

$$P_{\mathrm{in}} = P_{\mathrm{out}} + P_{\mathrm{c}} + P_{\mathrm{i}}, \tag{4.9}$$

where P_{out} is the sum of the mechanical power outputs of each motor, P_{c} is the sum of the copper losses of each motor, and P_{i} is the sum of the iron losses of each motor. The inverter loss is assumed to be negligible.

When the angular acceleration of each wheel is small, the torque T_j is proportional to the driving force. T_j can thus be expressed as

$$T_j \approx r F_j. \tag{4.10}$$

When the slip ratio λ_j is sufficiently small, ω_j is expressed as

$$\omega_j = \frac{V}{r(1-\lambda_j)} \approx \frac{V}{r}(1+\lambda_j). \tag{4.11}$$

Then, λ_j is expressed as

$$\lambda_j = \frac{F_j}{D_s' N_j(\dot{V})} = \frac{F_{\mathrm{all}}}{4D_s' N_j(\dot{V})}. \tag{4.12}$$

Therefore P_{out}, P_{c}, and P_{i} can be expressed as

$$\begin{aligned} P_{\mathrm{out}} &= 2\sum_{j=f,r} \omega_j T_j, \\ &\simeq V\frac{F_{\mathrm{all}}}{2}\sum_{j=f,r}\left(1+\frac{F_{\mathrm{all}}}{4D_s' N_j(\dot{V})}\right), \end{aligned} \tag{4.13}$$

$$P_{\mathrm{c}} = 2\sum_{j=f,r} R_j {i_{qj}}^2 = \frac{r^2}{8}{F_{\mathrm{all}}}^2 \sum_{j=f,r}\frac{R_j}{{K_{tj}}^2}, \tag{4.14}$$

$$P_{\mathrm{i}} = \frac{2V^2}{r^2}\sum_{j=f,r}\frac{{P_{nj}}^2}{R_{cj}}\left[\left(\frac{rL_{qj}F_{\mathrm{all}}}{4K_{tj}}\right)^2 + {\Psi_j}^2\right], \tag{4.15}$$

where R_j is the armature winding resistance of the motor, K_{tj} is the torque coefficient of the motor, P_{nj} is the number of pole pairs, L_{qj} is the q-axis inductance, and Ψ_j is the interlinkage magnetic flux.

The electrical angular speed of the motor ω_{ej} and the equivalent iron loss resistance R_{cj} are expressed as

$$\omega_{ej} = \frac{P_{nj}V}{r}, \tag{4.16}$$

$$\frac{1}{R_{cj}} = \frac{1}{R_{c0j}} + \frac{1}{R_{c1j}'\,|\omega_{ej}|}, \tag{4.17}$$

where the first and second terms on the right-hand side represent the eddy current loss and hysteresis loss, respectively.

Therefore, the inverter input power is expressed by V and F_{all} and is given by the following expression.

$$P_{\mathrm{in}}(V, F_{\mathrm{all}}) = P_{\mathrm{out}}(V, F_{\mathrm{all}}) + P_{\mathrm{c}}(F_{\mathrm{all}}) + P_{\mathrm{i}}(V, F_{\mathrm{all}}). \tag{4.18}$$

4.3 Optimization of Velocity Profile Considering Traffic Signal Information

In this section, we assumed that the autonomous driving systems are installed in the EVs, and thus proposed the READ. READ calculates optimal velocity profile, which minimizes the total energy consumption from the initial time t_0 to the final time t_f. In this study, we assume that the vehicle runs at $V_0 = 30.0\,\text{km/h}$ from the starting point $X_0 = 0.00\,\text{m}$ and that it stops with $V_f = 0.00\,\text{km/h}$ at the end point $X_f = 400\,\text{m}$ shown in Fig. 4.4a.

4.3.1 Signal Information Model

In this example, the traffic light is assumed to be installed at $X_1 = 200\,\text{m}$. The traffic signal information model is represented as

$$\mathrm{s}(t, X_1) = \begin{cases} 0 & \text{if } 0 \le t \le (1 - q_\mathrm{s})T_\mathrm{s} \\ 1 & \text{if } (1 - q_\mathrm{s})T_\mathrm{s} < t \le T_\mathrm{s}, \end{cases} \tag{4.19}$$

where $\mathrm{s}(t, X_1)$ is the traffic signal information at $X_1 = 200\,\text{m}$, T_s depicts one cycle, and q_s is the split. $\mathrm{s}(t, X_1) = 1$ implies a green light, and $\mathrm{s}(t, X_1) = 0$ indicates a red light. In this study, T_s and q_s are fixed at 80.0 s and 0.4, respectively in Fig. 4.4b.

4.3.2 Evaluation Function and Constraint Condition

The evaluation function and the constraint conditions are described as follows:

$$\text{min.}\;\; W_\mathrm{in} = \int_{t_0}^{t_f} P_\mathrm{in}(\boldsymbol{x}(t), u(t))dt \tag{4.20}$$

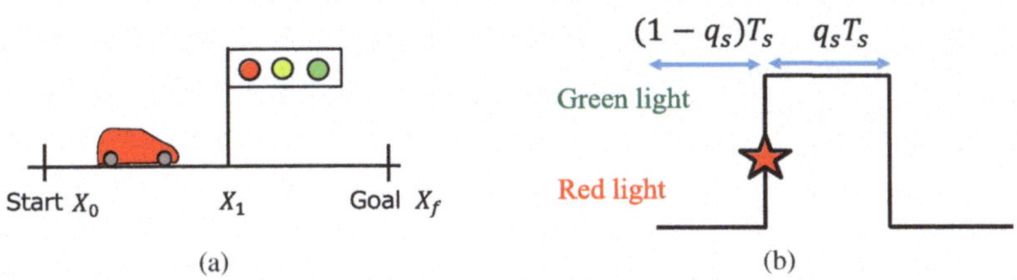

Fig. 4.4 Assumed scenario and traffic signal information [3]. (**a**) Assumed scenario. (**b**) Traffic signal information

$$
\begin{aligned}
s.t. \quad \dot{\boldsymbol{x}}(t) \quad &= \boldsymbol{f}(\boldsymbol{x}(t), u(t)), &(4.21)\\
\chi(\boldsymbol{x}(t_0)) &= \boldsymbol{x}(t_0) - \boldsymbol{x}_0 = \boldsymbol{0}, &(4.22)\\
\psi(\boldsymbol{x}(t_f)) &= \boldsymbol{x}(t_f) - \boldsymbol{x}_f = \boldsymbol{0}, &(4.23)\\
X(t) \quad &\leq X_1 \quad (t \leq (1 - q_s)T_s), &(4.24)\\
\boldsymbol{x}(t) = \left[V(t), X(t)\right]^{\mathrm{T}}&, u(t) = F_{\mathrm{all}}(t), &(4.25)
\end{aligned}
$$

where W_{in} is the total energy consumption, P_{in} is the inverter input power, V is the vehicle velocity, X is the distance traveled, and F_{all} is the total driving force. $\boldsymbol{x}_0$ and $\boldsymbol{x}_f$ are the initial and final state, respectively. In this study, the steepest descent method is used to calculate the velocity trajectory of the vehicle to solve the nonlinear optimization problem [1].

4.3.3 Comparison Conditions

To verify the performance of the proposed method, the obtained velocity profile is compared with two velocity profiles, which are regarded as the conventional methods. One is a simple optimized trajectory with traffic signal information. The other calculates the nonlinear optimization without the traffic signal information.

4.3.3.1 Conventional Profile 1: Constant Acceleration and Deceleration with Signal Information

Conventional profile 1 is composed of constant acceleration, constant speed, and constant deceleration as shown in (4.26).

$$
V(t, V_c) = \begin{cases} V_0 - a_1 t & \text{if } 0 \leq t \leq t_s, \\ V_{\mathrm{s}} + a_2(t - t_s) & \text{if } t_s \leq t \leq t_1, \\ V_{\mathrm{c}} & \text{if } t_1 \leq t \leq t_2, \\ V_{\mathrm{c}} - a_2(t - t_2) & \text{if } t_2 \leq t \leq t_f. \end{cases} \tag{4.26}
$$

This profile minimizes the energy consumption by autonomous driving using the traffic signal information, which means that X_1, T_s, and q_s are assumed to be well known. Then, the parameters of (4.26)–(4.32) are defined as

$$
t_s = (1 - q_s)T_s, \tag{4.27}
$$

$$
a_1 = \frac{V_0 t_s - X_1}{{t_s}^2/2}, \tag{4.28}
$$

$$V_s = -a_1 t_s + V_0, \tag{4.29}$$

$$a_2 = \frac{1}{2}\frac{-2V_c{}^2 - V_s{}^2 + 2V_s V_c}{X_f - X_1 - V_c(t_f - t_s)}, \tag{4.30}$$

$$t_1 = -\frac{V_s - V_c}{a_2} + t_s, \tag{4.31}$$

$$t_2 = t_f - \frac{V_c}{a_2}. \tag{4.32}$$

4.3.3.2 Conventional Profile 2: Optimized Velocity Profile Without Signal Information

Conventional profile 2 minimizes the energy consumption by autonomous driving considering the traffic signal phase. We assume that the vehicle obtains the signal phase not by ITS communication but through the in-vehicle camera. When the in-vehicle camera captures a red light, the vehicle uses the conventional READ without the time constraint strategy [4]. This profile is calculated to minimize the energy consumption from X_0 to X_1 without using the time constraint strategy. When the traffic signal light turns green, the velocity trajectory will change to the conventional READ with the time constraint strategy [4], which is calculated to minimize energy consumption from $X(t_s)$ to X_f up to t_f.

4.3.3.3 Proposed Profile: Optimized Velocity Profile with Signal Information

The proposed profile minimizes the energy consumption by autonomous driving, using the traffic signal information. The optimal trajectory of $\boldsymbol{x}(t)$ is obtained at the starting point to minimize the total energy consumption (4.20) with constraints (4.21)–(4.24).

4.4 Simulation Results

The results from the simulation are shown in Fig. 4.6. The red star in Fig. 4.6a indicates the point where the traffic signal changes from a red light to a green light.

To analyze the simulation results, the mechanical output P_{out} is categorized into the power stored as kinetic energy of vehicle mass P_{M}, the sum of the power stored as rotational energy of each wheel P_{J}, the loss incurred by driving resistance P_{R}, and the sum of the losses incurred by the slip condition of each wheel P_{S}. The integrated values of the above-mentioned parameters are described as

$$W_X = \int_{t_0}^{t_f} P_X(\boldsymbol{x}(t), u(t))dt, \tag{4.33}$$

where the subscript X represents "out," "M," "J," "R," "S," "c," and "i," as explained previously.

As shown in Fig. 4.6b, in conventional profile 1, the vehicle halts at the traffic signal. Subsequently, the vehicle accelerates to about 30 km/h near the traffic signal and eventually halts at the goal.

In conventional profile 2, when the in-vehicle camera captures a red light, the vehicle decelerates optimally to stop at the traffic signal to prevent the loss of kinetic energy caused by the driving resistance. As observable in Fig. 4.6e, this profile reduces the driving resistance loss more than conventional profile 1. When the traffic signal light turns green, the vehicle accelerates optimally to about 45 km/h and eventually halts at the end point to minimize the energy consumption from $X(t_s)$ to X_f. This prevents the vehicle from incurring motor loss and driving resistance loss. However, the total driving force and the velocity of conventional profile 2 are larger than those of conventional profile 1 and the proposed profile. Therefore, this results in an increase in the iron loss, copper loss, and driving resistance loss, which is more than that incurred by conventional profile 1, as shown in Fig. 4.6c–e. Therefore, the total energy loss of conventional profile 2 is the largest among the three velocity profiles, as shown in Fig. 4.6b.

In the proposed profile, when the traffic signal turns red, the vehicle reduces its speed at an optimal deceleration and travels at a low speed to the point of the traffic signal. The proposed profile incurs a lower driving resistance loss than conventional profile 1, as shown in Fig. 4.6e. When the traffic signal light turns green, the kinetic energy of proposed profile is larger than that of conventional profile 1 and 2. Hence, the proposed profile does not require the EV to accelerate more than that required for conventional profile 1 and 2. Therefore, the proposed profile reduces copper loss to a greater extent than conventional profile 1 and 2, as shown Fig. 4.6d.

The energy consumption of conventional profiles 1 and 2, and the proposed profile is 53.1, 61.6, and 44.3 kW, respectively, as shown in Fig. 4.6a and Table 4.3. The energy consumption of the proposed profile is lower than that of conventional profile 1 and 2 by 16.6% and 28.1%, respectively.

Figure 4.6b shows the total energy loss, which is composed of copper loss, iron loss, slip loss, and driving resistance loss. Copper loss is proportional to the square of total driving force, as shown in (4.14). Figure 4.6c illustrates that total driving

Table 4.3 Total energy consumption [kW] (simulation results)

Conventional 1	Conventional 2	Proposed
53.1	61.6	44.3

force of the proposed profile is the smallest among the three profiles from t_s to t_f. Therefore, copper losses incurred by the application of the proposed profile are reduced by 56.0% and 69.7% compared to those using conventional profile 1 and 2, respectively. Iron loss is proportional to the square of the total driving force and velocity, depicted by (4.15). Figure 4.5b and c shows that the total driving force and velocity of the proposed profile are the smallest ones among the three profiles, from t_s to t_f. Therefore, the iron losses incurred by the application of the proposed profile are reduced by 0.452% and 4.15% compared to those using conventional profile 1 and 2, respectively. The driving resistance is proportional to the square of velocity, as depicted by (4.4). Figure 4.6b shows that the velocity of the proposed profile is the smallest among the three profiles, from t_s to t_f. Therefore, the driving resistance loss incurred by the proposed profile is reduced by 2.05% and 10.8% compared to those using conventional profile 1 and 2, respectively.

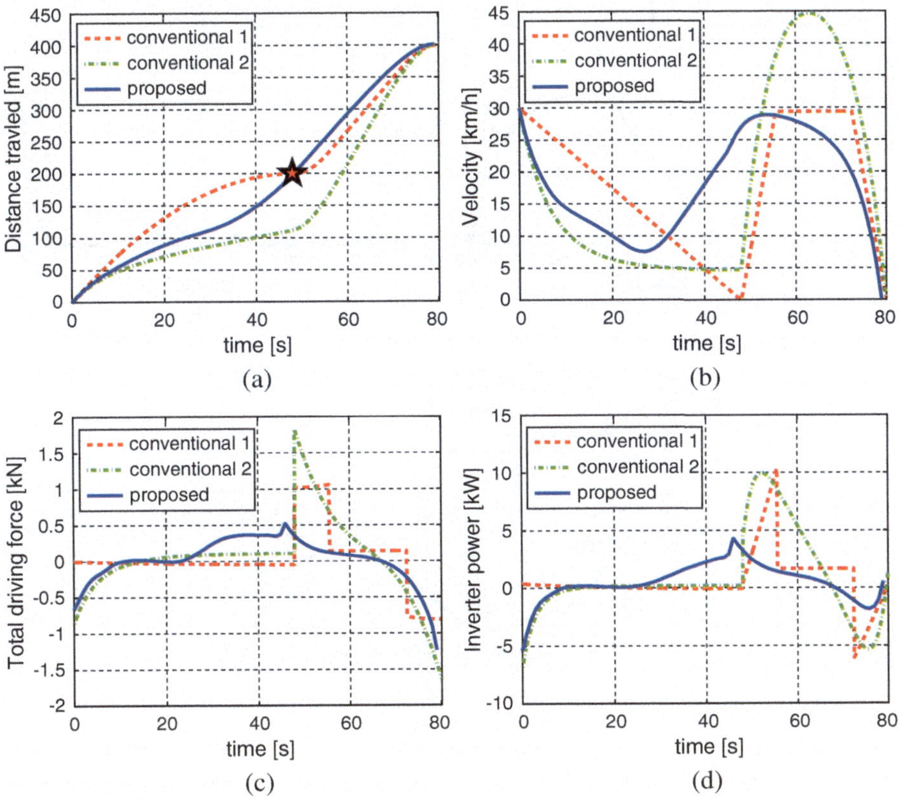

Fig. 4.5 Simulation results [3]. (**a**) Distance traveled. (**b**) Velocity. (**c**) Total driving force. (**d**) Inverter input power

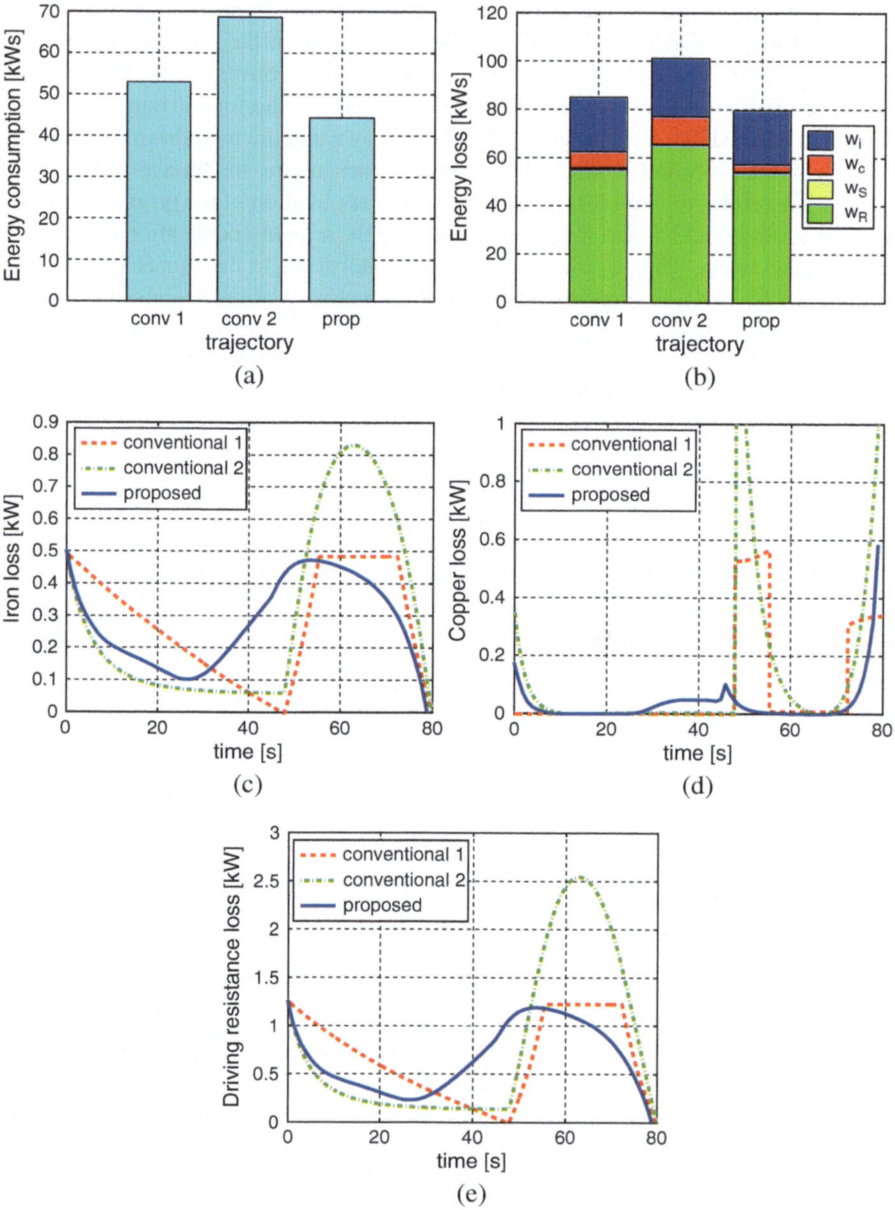

Fig. 4.6 Simulation results [3]. (**a**) Total energy consumption. (**b**) Total energy loss. (**c**) Iron loss. (**d**) Copper loss. (**e**) Driving resistance loss

4.5 Experimental Results

4.5.1 Control System

The vehicle velocity control system is designed to automatically control the EV's velocity. Figure 4.7 shows the block diagram of the control system, which is composed of a feedforward and a feedback controller. The front and rear torque reference $T_j{}^*$ is given as

$$T_j{}^* = rF_j{}^* + \frac{J_{\omega_j}\dot{V}^*}{r}(1+\lambda_j). \tag{4.34}$$

The second term on the right-hand side of the equation compensates for the inertia torque of the wheels. The vehicle velocity controller $C_{PI}(s)$ is a PI controller, and it is designed using the pole placement method. The closed-loop pole is placed at $-5\,\mathrm{rad/s}$.

4.5.2 Experimental Environment

Experiments were conducted on both the real car simulation bench test (RC-S) and road driving test, as shown in Fig. 4.8, under the same conditions as those of the simulations. RC-S developed by ONO Sokki Co., Ltd. can reproduce various travel scenarios without being influenced by the change of wind and road surface conditions. Vehicle velocity V and inverter input power P_{in} are calculated as

$$V = \frac{r}{4}\sum_{j=f,r}\sum_{i=l,r}\omega_{ij}, \tag{4.35}$$

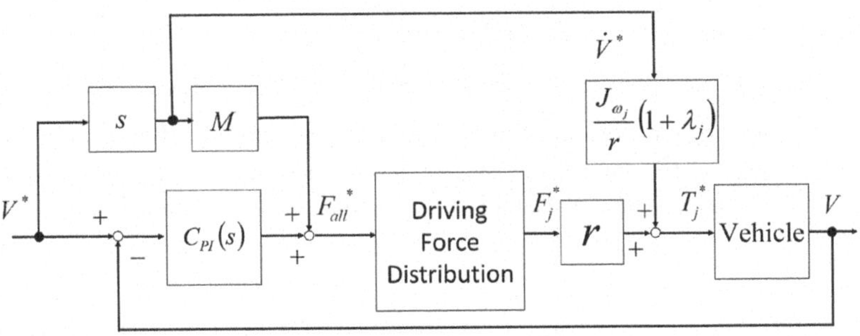

Fig. 4.7 Vehicle velocity control system [3]

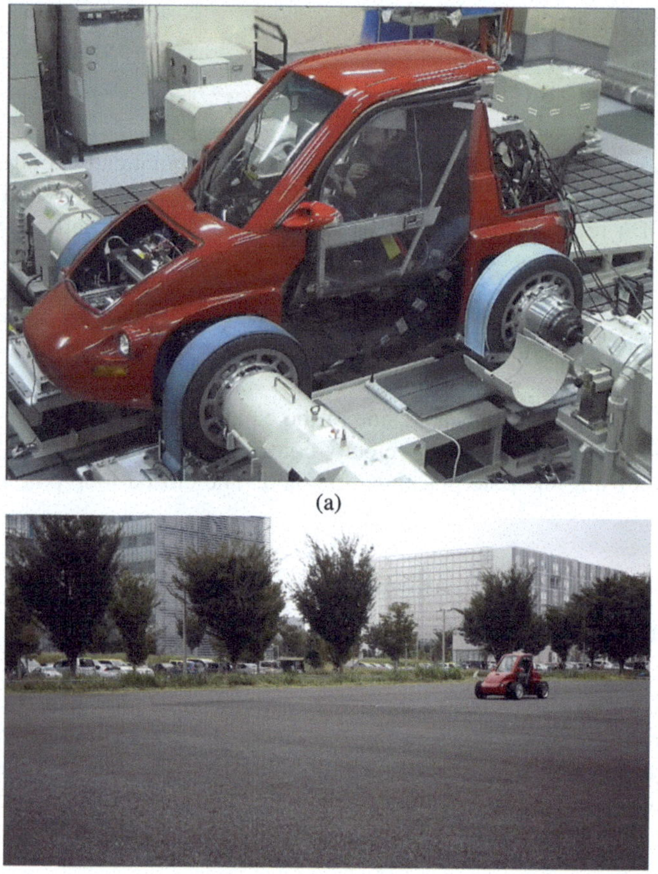

(a)

(b)

Fig. 4.8 Experiment environment [3]. (**a**) Real car simulation bench test (RC-S). (**b**) Driving test road

$$P_{\text{in}} = V_{\text{dc}} \sum_{j=f,r} I_{\text{dc}j}, \tag{4.36}$$

where the subscript i represents l or r (l denotes the left wheel and r denotes the right wheel). V_{dc} is the measured inverter input voltage and $I_{\text{dc}j}$ is the measured inverter input current. P_{in} also includes inverter losses incurred during power conversion.

4.5.3 Experiment Results of RC-S

Experiments were conducted on the RC-S under the same conditions as those in the simulations. Figure 4.9 depicts the experimental results. Figure 4.9d and Table 4.4 describe the total energy consumption, in terms of the average values and standard deviations of the experiments repeated four times.

The total energy consumption of conventional profile 1, conventional profile 2, and the proposed profile is 52.0, 57.5, and 49.4 kW, respectively. The average total energy consumption of the proposed profile is improved by 5.02 and 14.1%

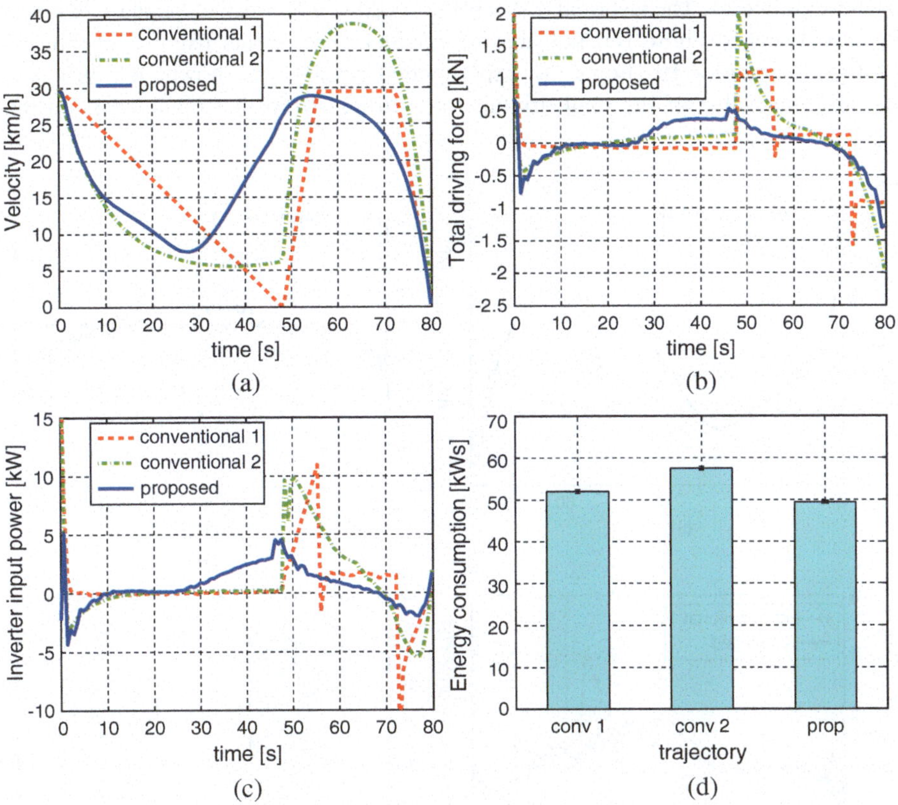

Fig. 4.9 Experiment results of RC-S [3]. (**a**) Velocity. (**b**) Total driving force. (**c**) Inverter input power. (**d**) Consumption energy

Table 4.4 Total energy consumption [kW] (average ± standard deviation)

	Conventional 1	Conventional 2	Proposed
RC-S	52.0 ± 0.423	57.5 ± 0.415	49.4 ± 0.343
Driving test	59.4 ± 5.19	N/A	50.5 ± 3.18

compared to that of conventional profile 1 and 2, respectively. The average total energy consumption in the experiments deviates from that obtained from the simulations because the neglect of the inverter loss affects the accuracy of the power flow model, which leads to an error. However, the trends of the experimental results are consistent with the simulation.

4.5.4 Experimental Results of the Driving Test

The experiments were conducted on the test driving road under the same condition as the simulations. The test road restricts the speed to a prescribed value of 30 km/h or less. We were, however, not able to conduct experiments using conventional profile 2. Figure 4.10 shows the experimental results. As shown in Fig. 4.10b, the absolute value of the total driving force differs from that obtained from the

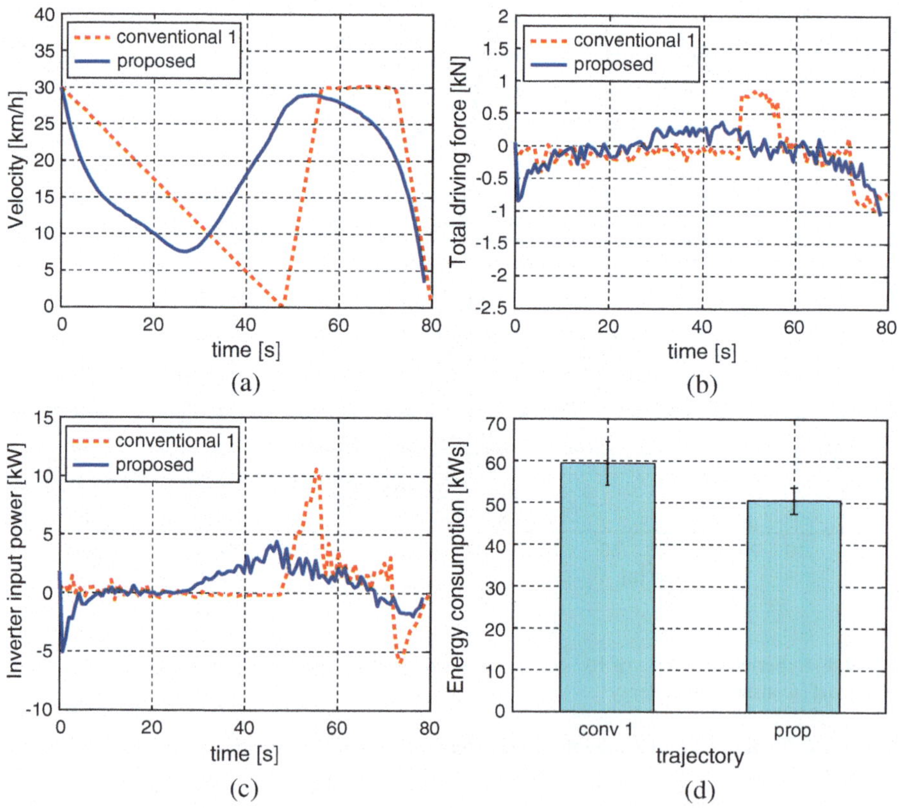

Fig. 4.10 Experiment results of driving test [3]. (**a**) Velocity. (**b**) Total driving force. (**c**) Inverter input power. (**d**) Consumption energy

simulations because of the modeling error of driving resistance. Figure 4.10d and Table 4.4 show the total energy consumption, which comprises the average values and standard deviations of the experiments that are repeated four times.

The total energy consumption of the conventional profile 1 and the proposed model are 59.4 and 50.5 kW, respectively. The average total energy consumption of the proposed profile is improved by 14.9% compared to that of conventional profile 1. The vehicle in the driving test was affected by road surface disturbance, whereas in the RC-S bench test, this was not the case. Moreover, the RC-S bench test gives the driving resistance using model (4.4). Therefore, the standard deviation of RC-S can be smaller than that of the driving test.

The experimental results verified that the proposed profile is superior to conventional profile 1 and 2 not only in the simulation test, but also in the road driving test in terms of the average total energy consumption.

4.6 Conclusion

In this chapter, we propose READ by considering the traffic signal from one traffic light. The proposed READ system was shown to reduce the energy consumption compared to conventional profile 1 and 2, through both simulation and experiments. The future works will include consideration of the space between the vehicle with READ and the vehicle in the front.

Acknowledgements This research was partly supported by the Industrial Technology Research Grant Program from New Energy and Industrial Technology Development Organization (NEDO) of Japan (number 05A48701d), Ministry of Education, Culture, Sports, Science and Technology grant (number 22246057 and 26249061), and the Core Research for Evolutional Science and Technology, Japan Science and Technology Agency (JST-CREST). The present study is a part of study being undertaken by the project team of JST-CREST named "Integrated Design of Local EMSs and their Aggregation Scenario Considering Energy Consumption Behaviors and Cooperative Use of Decentralized In-Vehicle Batteries."

References

1. J.T. Betts, *Practical Methods for Optimal Control Using Nonlinear Programming*, 2nd ed., SIAM, Philadelphia (2010)
2. H. Fujimoto, S. Harada, Model-based range extension control system for electric vehicles with front and rear driving–braking force distributions. IEEE Trans. Ind. Electron. **62**(5), 3245–3254 (2015)
3. H. Fujimoto, H. Yoshida, D. Kawano, Y. Goto, M. Tsuchimoto, K. Sato, Bench test of range extension autonomous driving for electric vehicles based on optimization of velocity profile considering traffic signal information, in *International Electric Vehicle Technology Conference and Automotive Power Electronics Japan*, Yokohama (2016)
4. S. Harada, H. Fujimoto, Range extension control system based on optimization of acceleration-deceleration profile and front and rear driving-braking force distribution, in *Multi-symposium on Control Systems* (in Japanese), (2014) pp. 6F1–4

5. H. Hijikata, K. Akatsu, Y. Miyama, H. Arita, A. Daikoku, Suppression control method for iron loss of MATRIX motor under flux weakening utilizing individual winding current control, in *The 7th International Power Electronics Conference*, IPEC-Hiroshima 2014, ECCE Asia (2014), pp. 2673–2678
6. Y. Hori, Future vehicle driven by electricity and control–research on four–wheel–motored "UOT Electric March II". IEEE Trans. Ind. Electron. **51**(5), 954–962 (2004)
7. Y. Ikezawa, H. Fujimoto, Y. Hori, Range extension autonomous driving for electric vehicles based on optimal vehicle velocity profile generation and front–rear driving–braking force distribution with time constraint, in *The 1st IEEJ International Workshop on Sensing Actuation and Motion Control (SAMCON)* (2015), pp. 1–6
8. M.A.S. Kamal, M. Mukai, J. Murata, T. Kawabe, Model predictive control of vehicles on urban roads for improved fuel economy. IEEE Trans. Control Syst. Technol. **21**(3), 831–841 (2013)
9. H. Kimura, H. Otake, H. Oguri, M. Kanematsu, Field evaluation test results of green wave advisory system (in Japanese), in *JSAE Annual Congress (Autumn), Inc.*, No. 87–12 (2012), pp. 15–18
10. T. Kobayashi, E. Katsuyama, H. Sugiura, E. Ono, M. Yamamoto, Study on driving force distribution and power consumption in cornering: formulation for steady state cornering and validation using EV. JSAE Trans. **45**(2), 309–314 (2014) (in Japanese)
11. R. Nakajima, R. Koyama, D. Matsuoka, Y. Kano, M. Abe, The effects of tire force distribution control on reducing tire slip energy dissipation (in Japanese), in *2013 JSAE Annual Congress (Spring)*, No. 30 (2013), pp. 21–24
12. O. Nishihara, T. Kumazawa, Electric power conservation by driving/braking force distribution in electric vehicle (in Japanese), in *Dynamics and Design Conference*(2010), pp. 330–1–300–6
13. H.B. Pacejka, E. Bakker, The magic formula tyre model. Veh. Syst. Dyn. Int. J. Veh. Mech. Mobil. **21**(1), 1–18 (1992)
14. X. Wu, X. He, G. Yu, A. Harmandayan, Y. Wang, Energy-optimal speed control for electric vehicles on signalized arterials. IEEE Trans. Intell. Transp. Syst. 1–11 (2015)

Part II
Synthesis of Distributed Energy Management Systems Based on Aggregation of Local EMSs and Vehicles

Chapter 5
Real-Time Pricing and Decentralized Optimization Strategy for Power Flow Balancing in EV/PHV Storage Management

Kenji Hirata

5.1 Introduction

Control design problems usually consider regulation and tracking of system states with respect to known set-points or trajectories. Optimal set-points or trajectories are typically given or determined a priori by solving an appropriate optimization problem. The optimization problem may incorporate economical efficiency, security limits of the plant, and operating constraints associated with the steady-state operation of the entire system. The control system may be required to follow a time-dependent demand, and the optimization problem should be solved again to reflect new demand conditions and update optimal set-points or trajectories. Considering a large-scale system consisting of a large number of components or agents, an appropriate optimization problem becomes large, and solving the problem becomes a non-immediate task. A promising approach for overcoming this difficulty is a decentralized mechanism, where each agent determines its optimal set-point, while the solution should satisfy specified operating constraints for social benefits.

Motivated by the decentralized charging/discharging management for EV storage, this chapter investigates the optimal operation problem with steady-state constraints under decentralized decision-making, where the steady-state constraints corresponds to the power flow balance between the prediction-based power supply and actual demands. Each agent, a dynamic component of the entire system, is allowed to determine its own optimal set-point according to individual profit. An independent entity, referred to as the utility in this chapter, aspires to realize a socially optimal solution that fulfills steady-state operating constraints. The individual decision-making of the agents is unlikely to satisfy the steady-state

K. Hirata (✉)
Graduate School of Science and Engineering for Research, University of Toyama, Toyama, Japan
e-mail: hirata@eng.u-toyama.ac.jp

T. Suzuki et al. (eds.), *Design and Analysis of Distributed Energy Management Systems*, Power Electronics and Power Systems,
https://doi.org/10.1007/978-3-030-33672-1_5

constraints. To align the individual decision-making of each agent with a socially optimal solution, the utility is allowed to provide an additional price, which conceptually represents tax to the agent or subsidy from the community. Each agent will participate in the community through optimization of the individual profit including the provided additional price.

We propose a real-time pricing strategy of the utility. The resulting closed-loop system eventually realizes tracking in the socially optimized solution, where there is no need to solve a large-scale optimization problem nor to consider iterative computations to determine the price. The proposed real-time pricing strategy is heuristic, since it combines the gradient method to determine the price, which is conceptually related to Uzawa algorithm [1], individual decision-making, and the dynamics of the agents. However, we show that the resulting closed-loop system is stable at least locally, if the step-size parameter in the real-time pricing strategy is sufficiently small.

The proposed real-time pricing and decentralized decision-making methodology is applied to the decentralized charging/discharging management problem for EV storage. The mechanism to determine the pricing signal uses the amount of net imbalance of the power flow, thus the individual power consumption of each vehicle does not need to be reported. Moreover, the mechanism does not require to know the number of vehicles that are temporarily connected. It allows a plug-and-play type operation and is useful in the vehicle charging/discharging management, since the number of available vehicles is usually unknown. The numerical case studies illustrate the effectiveness of the proposed real-time pricing and decentralized decision-making methodology.

5.2 Real-Time Pricing and Decentralized Optimization Leading to Optimal Operation

5.2.1 Dynamics of Agent

We consider a group of n agents, G_i, $i \in N = \{1, \ldots, n\}$. Each agent G_i has a group of neighbors, and we denote by $N_i \subset N$ the set of indices of the neighbors of G_i. The agent G_i has interactions with its neighbors and is represented by an equation of the form

$$\dot{x}_i(t) = f_i(x_i(t), \zeta_i(t), w_i(t), r_i(t)), \tag{5.1a}$$

$$z_i(t) = g_i(x_i(t), \zeta_i(t), w_i(t)), \tag{5.1b}$$

where $x_i \in X_i \subset \mathbb{R}^{n_i}$ denotes the state, $r_i \in R^{m_i}$ denotes the set-point, $z_i \in R^{m_i}$ denotes the output to be tracked to r_i, $w_i \in W_i \subset \mathbb{R}^{m_{wi}}$ denotes an exogenous input, and $\zeta_i \in \mathbb{R}^{m^i}$ represents interactions between the agents with $\xi_i \in \prod_{j \in N_i} X_i$ and

$$\zeta_i(t) = k_i(\xi_i(t)). \tag{5.1c}$$

We assume that each function in (5.1) is differentiable.

5.2.2 Decentralized Determination of Optimal Set-Point

We suppose that each agent G_i is allowed to select its own set-point r_i. We set an optimization problem

$$\min_{r_i} \quad J_i(w_i; r_i) \tag{5.2a}$$

$$\text{subject to} \quad h_{ij}(w_i; r_i) \le 0, \qquad j \in T_i = \{1, 2, \ldots, t_i\}, \tag{5.2b}$$

for each G_i, $i \in N$ and suppose that an economically optimal set-point is given as its optimal solution. For each given $w_i \in W_i$, we suppose that $J_i(w_i; \cdot) : \mathbb{R}^{m_i} \to \mathbb{R}$ is strictly convex and differentiable, and $h_{ij}(w_i; \cdot) : \mathbb{R}^{m_i} \to \mathbb{R}$ is convex and differentiable. The optimal solution to (5.2) is denoted by $r_i^\sharp$.

An independent entity, referred to as the utility in the remainder of this chapter, likewise attempts to incorporate the economic efficiency of the agents. However, its most important priority is to fulfill operating constraints at the steady-state such as $L(w)z + \ell_w(w) = 0$, where, we set $z = [\, z_i \,]_{i\in N}$, $w = [\, w_i \,]_{i\in\{0\}\cup N} \in W = W_0 \times \prod_{i\in N} W_i$, and $w_0 \in W_0 \subset \mathbb{R}^{m_{w0}}$ denotes an additional exogenous input to represent the steady-state constraints. The utility attempts to determine the socially optimal solution as the optimal solution to the following problem:

$$\min_{r} \quad \sum_{i\in N} J_i(w_i; r_i) \tag{5.3a}$$

$$\text{subject to} \quad h_{ij}(w_i; r_i) \le 0, \qquad i \in N, \quad j \in T_i, \tag{5.3b}$$

$$L(w)r + \ell_w(w) = 0, \tag{5.3c}$$

where $r = [r_i]_{i\in N} \in \mathbb{R}^m$ and $m = \sum_{i\in N} m_i$. The matrices $L(w) = [L_1(w) \;\; L_2(w) \;\; \cdots \;\; L_n(w)] \in \mathbb{R}^{\ell\times m}$, $L_i : W \to \mathbb{R}^{\ell\times m_i}$, $i \in N$ and $\ell_w : W \to \mathbb{R}^\ell$ define equality constraints that the utility aspires to satisfy. The optimal solution to (5.3) is denoted by r_i^*.

The decentralized decision-making $r_i^\sharp$ of each agent in (5.2) is unlikely to be aligned with the socially optimal solution r_i^* to (5.3). To align the individual decision-making of each agent with the socially optimal solution, we suppose that the utility is allowed to provide an additional price $p_i \in \mathbb{R}^{m_i}$, $i \in N$, or, in other words, charge a tax to the agent or receive a subsidy from the community. Each agent G_i participates in the community through the optimization of the following problem:

$$\min_{r_i} \quad J_i(w_i; r_i) + p_i^{\mathrm{T}} r_i \tag{5.4a}$$

$$\text{subject to} \quad h_{ij}(w_i; r_i) \le 0, \qquad j \in T_i. \tag{5.4b}$$

For each given p_i, The optimal solution to (5.4) is denoted by $r_i^{\flat}(p_i)$.

We suppose that the value of the net imbalance $L(w)z(t) + \ell_w(w)$ is available to the utility as a real-time measurement and consider the feedback interactions between the agents G_i, $i \in N$ and the utility. The utility tries to determine the price $p_i(t)$ in real-time, and the agent G_i determines its own set-point $r_i(t) = r^{\flat}(p_i(t))$ as the solution to (5.4) in real-time. This chapter considers the design of the real-time pricing strategy for the utility that determines $p_i(t)$ using the net imbalance $L(w)z(t) + \ell_w(w)$ and eventually realizes $r_i^{\flat}(p_i(t)) \to r_i^*$ for all $i \in N$.

We list the standing assumptions, which are instrumental in the remaining of this chapter.

Assumption 5.1 For each $w \in W$, matrix $L(w)$ has a full row rank. □

Assumption 5.1 implies that the representation of the equality constraints in (5.3c) includes no redundant constraint.

Assumption 5.2 For each $w \in W$, the constraints in (5.3b) and (5.3c) satisfy Slater's constraint qualification condition. □

Slater's constraint qualification condition implies that a strong duality holds for the problem (5.3) [2], and that the KKT (Karush–Kuhn–Tucker) conditions are necessary and sufficient conditions for optimality.

Assumption 5.3 For each $w \in W$, there exists an optimal solution to (5.3). □

Due to the strict convexity of each J_i, the optimal solution to (5.3) is unique for each $w \in W$, if it exists.

Assumption 5.4 Let $w \in W$ and r be given constants:

1. There exits an equilibrium point x that satisfies

$$0 = f_i(x_i, k_i(\xi_i), w_i, r_i),$$

   for all $i \in N$.
2. The equilibrium point x satisfies

$$r_i = g_i(x_i, k_i(\xi_i), w_i), \tag{5.5}$$

   for all $i \in N$.
3. There exists a linearization of (5.1) around (x, r)

$$\delta \dot{x}_i = a_i(w)\delta x_i + \hat{a}_i(w)\delta\xi_i + b_i(w)\delta r_i, \tag{5.6a}$$

$$\delta z_i = c_i(w)\delta x_i + \hat{c}_i(w)\delta\xi_i, \qquad i \in N, \tag{5.6b}$$

where $a_i(w) = \partial f_i/\partial x_i$, $\hat{a}_i(w) = \partial(f_i k_i)/\partial\xi_i$, $b_i(w) = \partial f_i/\partial r_i$, $c_i(w) = \partial g_i/\partial x_i$, and $\hat{c}_i(w) = \partial(g_i k_i)/\partial\xi_i$, respectively.

4. Considering the matrices $A(w)$, $B(w)$, and $C(w)$ that arise from the local linearization (5.6) of (5.1), the matrix $A(w)$ is a Hurwitz. □

We suppose that (5.1) represents a closed-loop system that consists of a plant to be controlled and a local feedback controller, which has already been designed. Assumption 5.4 requires that this closed-loop system results in error-free steady-state tracking, $\lim_{t\to\infty} z_i(t) = r_i$, at least locally.

5.2.3 Real-Time Pricing Strategy

This section proposes a real-time pricing strategy for the utility, which uses the net imbalance of the constraints (5.3c). The agent will determine the set-point $r^{\flat}(p_i)$ according to (5.4) applying the provided price p_i. Section 5.2.3.1 investigates a pricing strategy in the steady-state. The dynamic (real-time) case is considered in Sect. 5.2.3.2.

5.2.3.1 Pricing Strategy in Steady-State

This section investigates steady-state optimality. We provide a pricing strategy for the utility that aligns the distributed decision-making $r_i^{\flat}$ of the agent in (5.4) with the socially optimal solution r_i^* to (5.3).

We start with the KKT conditions of (5.3).

$$\frac{\partial J_i(w_i; r_i)}{\partial r_i} + L_i^{\mathrm{T}}(w)\lambda + \sum_{j\in T_i} \frac{\partial h_{ij}(w_i; r_i)}{\partial r_i}\mu_{ij} = 0, \tag{5.7a}$$

$$h_{ij}(w_i; r_i) \le 0, \quad \mu_{ij} \ge 0, \quad h_{ij}(w_i; r_i) \times \mu_{ij} = 0, \qquad i \in N, \quad j \in T_i, \tag{5.7b}$$

$$L(w)r + \ell_w(w) = 0. \tag{5.7c}$$

From Assumption 5.2, a strong duality holds for (5.3), and the optimal solution to (5.3) can be determined from the conditions in (5.7).

The following Lemma 5.1 states that, at least in the steady-state, the utility can choose the optimal price p_i^* that realizes $r_i^{\flat}(p_i^*) = r_i^*$ for all $i \in N$.

Lemma 5.1 *Let $w \in W$ be given, and let us denote the optimal solution to* (5.3) *or, equivalently, denote variables that satisfy the KKT conditions in* (5.7) *as r_i^*, μ_{ij}^*, $i \in N$, $j \in T_i$ and λ^*. If the utility provides the price, p_i, according to the pricing strategy $p_i = L_i^{\mathrm{T}}(w)\lambda \in \mathbb{R}^{m_i}$, $i \in N$, then we obtain $r_i^{\flat}(p_i^*) = r_i^{\flat}(L_i^{\mathrm{T}}(w)\lambda^*) = r_i^*$.*
□

This basic result has been considered in literature, e.g., [6–8] and it is known as the concept of shadow price in the optimization literature [2].

5.2.3.2 Gradient Based Real-Time Pricing Strategy

The utility needs to solve the optimization problem (5.3) or the KKT conditions in (5.7) to determine the optimal price p_i^*. However, we do not consider this type of centralized operation. This section proposes a real-time pricing strategy for the utility based on the net imbalance of the constraints (5.3c).

We start with the dual problem of (5.3)

$$\max_{\lambda} \min_{\substack{r \\ h_{ij}(w_i;r_i)\le 0 \\ i\in N \quad j\in T_i}} \sum_{i\in N} J_i(w_i;r_i) + \lambda^{\mathrm{T}}(L(w)r + \ell_w(w)).$$

We suppose that r_i will be determined as $r_i^{\flat}$ by each agent G_i. By substituting it, we have

$$\max_{\lambda} \sum_{i\in N} J_i(w_i;r_i^{\flat}) + \lambda^{\mathrm{T}}(L(w)r^{\flat} + \ell_w(w)),$$

where $r^{\flat} = [\, r_i^{\flat} \,]_{i\in N}$. Applying the gradient method to this maximization problem leads to

$$\frac{d\lambda}{d\tau} = \epsilon(L(w)r^{\flat} + \ell_w(w)), \qquad \epsilon > 0,$$

where the constant $\epsilon > 0$ is the step-size parameter.

We replace $r^{\flat}$ by its corresponding output $z(t)$ in the above equation and combine the result in Lemma 5.1. Hence, a heuristic real-time pricing strategy of the utility is given as

$$\dot{\lambda}(t) = \epsilon(L(w)z(t) + \ell_w(w)), \qquad \epsilon > 0, \tag{5.8a}$$

$$p_i(t) = L_i^{\mathrm{T}}(w)\lambda(t), \qquad i \in N. \tag{5.8b}$$

Figure 5.1 shows a schematic block diagram of the proposed closed-loop system that includes gradient based pricing strategy (5.8), decentralized decision-making (5.4) of each agent G_i, and coupled agents dynamics (5.1). In the proposed closed-loop system, solving a large-scale optimization problem is not necessary nor is considering iterative calculations to determine the price. The utility only requires to know the amount of net imbalance $L(w)z(t) + \ell_w(w)$ and does not need to know each $z_i(t)$. Moreover, any specific properties of the agent G_i, such as functions f_i or J_i, do not need to be reported to the utility. The real-time pricing strategy (5.8) does

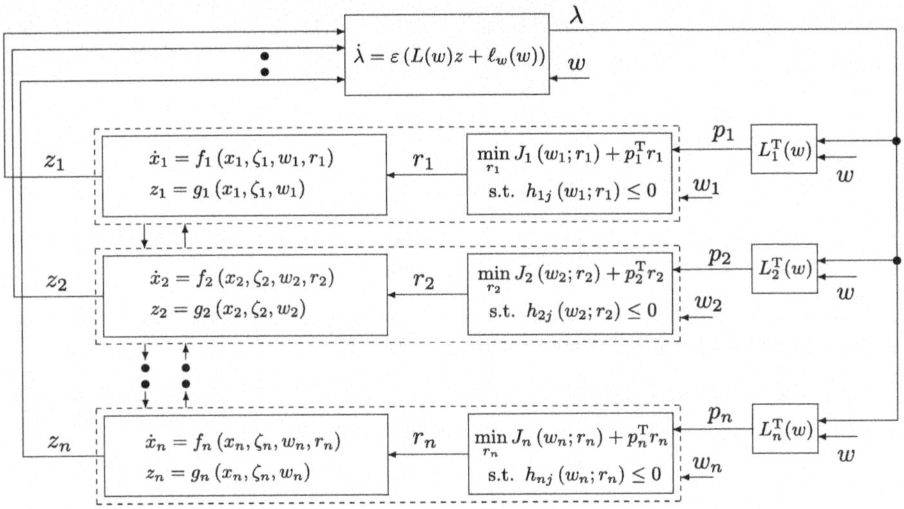

Fig. 5.1 A schematic block diagram of the decentralized optimization and integrate pricing strategy mechanisms, where the utility uses the net imbalance $L(w)z(t) + \ell_w(w)$ to determine the price

not depend on the number of agents that are temporally connected. This is useful in the problem of vehicle charging/discharging management, since a precise prediction of vehicle travels may be difficult.

5.2.4 Stability Analysis

This section investigates the stability of the closed-loop system in Fig. 5.1. The pricing strategy (5.8) heuristically combines the gradient method for the maximization problem to determine λ and dynamics (5.1) of each agent G_i, thus the stability of the resulting closed-loop system is not clear. Furthermore, the closed-loop system includes the optimization problem (5.4), and the stability analysis does not seem to be straightforward; hence, global stability analysis may not be easy. To capture the local dynamics around the equilibrium point, we first replace each optimization problem by its gradient in Sect. 5.2.4.1. We need to investigate the uniqueness of the equilibrium point of this local dynamics, which will be considered in Sect. 5.2.4.2. In Sect. 5.2.4.3, we show that the resulting closed-loop system in Fig. 5.1 is stable at least locally for sufficiently small $\epsilon > 0$.

Definition 5.1 Let $w \in W$ be a given constant. A triple (x, λ, r) is an equilibrium point of the closed-loop system in Fig. 5.1 correspondingto w, if

$$0 = f_i(x_i, k_i(\xi_i), w_i, r_i), \tag{5.9a}$$

$$0 = \epsilon(L(w)g(x, w) + \ell_w(w)), \tag{5.9b}$$

$$r_i = r_i^\flat(L_i^\mathrm{T}(w)\lambda) \tag{5.9c}$$

hold for all $i \in N$, where $g(x, w) = [g_i(x_i, k_i(\xi_i), w_i)]_{i \in N}$. □

Note that x and λ are a part of the state, whereas r is not. The following lemma is a conclusion from Assumptions 5.3, 5.4-1, and 5.4-2.

Lemma 5.2 *For each given constant $w \in W$, the closed-loop system in Fig. 5.1 has an equilibrium point (x, λ, r) that satisfies* (5.5).

Let r^* be an optimal solution to (5.3) for a given $w \in W$. x^* and λ^* exist such that the triple (x^*, λ^*, r^*) can be an equilibrium point, and satisfy (5.5).

5.2.4.1 Local Behavior of Decentralized Decision-Making

We capture (5.4) as a mapping from the price p_i to the set-point r_i, and investigate the gradient of this mapping. Let $w \in W$ be given, and r_i^*, μ_{ij}^*, $i \in N$, $j \in T_i$ and λ^* denote the corresponding optimal solution to (5.3) or, equivalently, the variables that satisfy the KKT conditions in (5.7). We set $p_i^* = L_i^\mathrm{T}(w)\lambda^*$, $i \in N$ and consider the optimal solution $r_i^\flat(p_i^*)$ to (5.4). We define $I_i(p_i^*) \subset T_i$ for each $i \in N$, possibly $I_i(p_i^*) = \varnothing$, as $I_i(p_i^*) = \{j \in T_i | \ h_{ij}(w_i; r_i^\flat(p_i^*)) = 0\}$. The set $I_i(p_i^*)$ contains all index $j \in T_i$ such that the inequality constraint in (5.4b) becomes active with the optimal solution $r_i^\flat(p^*)$, and the optimality conditions of the problem (5.4) are given as

$$\frac{\partial J_i(w_i; r_i)}{\partial r_i} + p_i^* + \sum_{j \in I_i(p_i^*)} \frac{\partial h_{ij}(w_i; r_i)}{\partial r_i}\mu_{ij} = 0, \tag{5.10a}$$

$$\mu_{ij} \geq 0 \quad h_{ij}(w_i; r_i) = 0, \quad j \in I_i(p_i^*). \tag{5.10b}$$

We now consider the following assumption.

Assumption 5.5 There exists $\delta_i > 0$, $i \in N$ such that $I_i(p_i^*) = I_i(p_i^* + \delta p_i)$ holds for all $\delta p_i \in U_{p_i}$, where $U_{p_i} = \{\delta p_i \in \mathbb{R}^{m_i} | \ \|\delta p_i\| \leq \delta_i\}$. □

Since $\delta_i > 0$, $i \in N$ are possibly sufficiently small, and the assumption considers the case such that the constraint h_{ij} that is active with p_i^* still remains active with $p^* + \delta p_i$. Essentially, the assumption holds unless the complementary conditions in (5.10b) are valid with both $\mu_{ij} = 0$ and $h_{ij} = 0$. Under Assumption 5.5, the conditions in (5.10) hold identically for all $p_i \in \{p_i^*\} + U_{p_i}$, and hence we consider the derivative of (5.10) and obtain

$$\frac{\partial r_i}{\partial p_i}\left(\frac{\partial^2 J_i}{\partial r_i^2} + \sum_{j\in I_i(p_i^*)} \mu_{ij}\frac{\partial^2 h_{ij}}{\partial r_i^2}\right) + \sum_{j\in I_i(p_i^*)} \frac{\partial \mu_{ij}}{\partial p_i}\left(\frac{\partial h_{ij}}{\partial r_i}\right)^{\mathrm{T}} = -I_{m_i}, \tag{5.11a}$$

$$\frac{\partial r_i}{\partial p_i}\frac{\partial h_{ij}}{\partial r_i} = 0 \quad j \in I_i(p_i^*). \tag{5.11b}$$

We set $L_i(w_i; r_i, \mu_{ij}) = J_i(w_i; r_i) + p_i^{\mathrm{T}} r_i + \sum_{j\in I_i(p_i^*)} h_{ij}(w_i; r_i)\mu_{ij}$ and, by multiplying the matrix $(\partial r/\partial p)^{\mathrm{T}}$ from the right to (5.11a), we have that

$$\frac{\partial r_i}{\partial p_i}\frac{\partial^2 L_i}{\partial r_i^2}\left(\frac{\partial r_i}{\partial p_i}\right)^{\mathrm{T}} + \left(\frac{\partial r_i}{\partial p_i}\right)^{\mathrm{T}} = 0.$$

From the necessary condition for the optimality, we have $\partial^2 L_i/\partial r_i^2 > 0$, and it concludes that

$$\frac{\partial r_i}{\partial p_i} = -\frac{\partial r_i}{\partial p_i}\frac{\partial^2 L_i}{\partial r_i^2}\left(\frac{\partial r_i}{\partial p_i}\right)^{\mathrm{T}} \leq 0.$$

We set $E_i = -(\partial r_i/\partial p_i)^{\mathrm{T}} = -(\partial r_i/\partial p_i) \geq 0$ and have

$$r_i^\flat(p_i^* + \delta p_i) = r_i^\flat(p_i^*) - E_i\delta p_i + \mathscr{O}(\|\delta p_i\|^2),$$

for a small deviation δp_i from p_i^*. We further set $\delta r_i = -E_i\delta p_i$ to denote the small deviation from $r_i^* = r^\flat(p_i^*)$ due to the small deviation δp_i from p_i^*. The gradient E_i represents the local behavior around the equilibrium point (x, λ, r) corresponding to w, which eventually depends on a given $w \in W$. We denote this by $E_i(w)$ in the remainder of this chapter.

5.2.4.2 Uniqueness of the Equilibrium Point in Local Dynamics

Let $w \in W$ be given, and let a triple (x, λ, r) be an equilibrium point of the closed-loop system in Fig. 5.1 corresponding to w. The triple (x, λ, r) is likewise an equilibrium point of the closed-loop dynamics, where optimization problems (5.4) are replaced by the gradients $E_i(w)$. However, this may not represent a unique equilibrium point. We start with the following definition.

Definition 5.2 Let $w \in W$ be a given constant, and let a triple (x, λ, r) be an equilibrium point in Lemma 5.2. The triple (x, λ, r) is said to be a unique equilibrium point in the local behavior, if

$$0 = f_i(x_i + \delta x_i, k_i(\xi_i + \delta \xi_i), w_i, r_i + \delta r_i), \tag{5.12a}$$

$$0 = \epsilon(L(w)g(x + \delta x, w) + \ell_w(w)), \tag{5.12b}$$

$$r_i + \delta r_i = r_i^\flat(L_i^\mathrm{T}(w)\lambda) - E_i(w)L_i^\mathrm{T}(w)\delta\lambda, \tag{5.12c}$$

for all $i \in N$ imply $\delta x = 0$, $\delta\lambda = 0$ and $\delta r = 0$. □

Let the triple (x, λ, r) be a unique equilibrium point in local behavior. If one considers the linearized dynamics of the closed-loop system around (x, λ, r), then $(0, 0, 0)$ is a unique equilibrium point of the linear dynamics.

Lemma 5.3 *Let $w \in W$ be a given constant, and let the triple (x, λ, r) be an equilibrium point in Lemma 5.2. We set $E(w) = \text{block diag}(E_1(w), E_2(w), \ldots, E_n(w))$. The following two statements are equivalent:*

1. *The triple (x, λ, r) is a unique equilibrium point in local behavior.*
2. *The matrix $L(w)E(w)L^\mathrm{T}(w)$ is non-singular.* □

The matrix $L(w)$ has a full row rank, rank $L(w) = \ell$, but each matrix $E_i(w)$ is only positive semi-definite, $E_i(w) \geq 0$, and there is a possibility that $L(w)E(w)L^\mathrm{T}(w)$ becomes singular.

5.2.4.3 Local Stability of the Closed-Loop System

Let $w \in W$ be given, and let a triple (x, λ, r) denote an equilibrium point in Lemma 5.2. As in Assumptions 5.4-3 and 5.4-4, around (x, λ, r), we consider the linearized dynamics of each agent and the matrices $A(w)$, $B(w)$, and $C(w)$ that arise from the local linearization (5.6) of (5.1). Together with $\delta\dot{\lambda}(t) = \epsilon L(w)\delta z(t)$, $\delta p = L^\mathrm{T}(w)\delta\lambda$ and $\delta r = -E(w)\delta p$, the linearized dynamics of the closed-loop system in Fig. 5.1 is given by

$$\begin{bmatrix} \delta\dot{x} \\ \delta\dot{\lambda} \end{bmatrix} = \begin{bmatrix} A(w) & -B(w)E(w)L^\mathrm{T}(w) \\ \epsilon L(w)C(w) & 0 \end{bmatrix} \begin{bmatrix} \delta x \\ \delta\lambda \end{bmatrix}, \tag{5.13}$$

where $\delta x = [\,\delta x_i\,]_{i \in N}$ and $\delta\lambda$ denote small deviation from x and λ, respectively. We immediately obtain the following local stability result.

Lemma 5.4 *Let $w \in W$ be a given constant and a triple (x, λ, r) be an equilibrium point in Lemma 5.2. The triple (x, λ, r) is a stable equilibrium point of the closed-loop system in Fig. 5.1 at least locally, if the linear system in* (5.13) *is stable.* □

If the matrix $L(w)E(w)L^\mathrm{T}(w)$ is singular, the linear system in (5.13) necessarily has a zero-eigenvalue and hence is not stable. Thus, $L(w)E(w)L^\mathrm{T}(w) > 0$ is a necessary condition for stability of the linear system in (5.13). Lemma 5.4 states general local stability conditions. However, we may consider a large network with a huge number of agents, and a numerical approach for checking the stability of a large size matrix in (5.13) may become difficult. A structural stability condition in the following Theorem 5.1 may be preferred.

Theorem 5.1 *Let $w \in W$ be a given constant, and let a triple (x, λ, r) be an equilibrium point in Lemma 5.2. Suppose that the matrix $L(w)E(w)L^{\mathrm{T}}(w)$ is non-singular. The triple (x, λ, r) is a stable equilibrium point of the closed-loop system in Fig. 5.1 at least locally, if $\epsilon > 0$ in (5.8) is sufficiently small.* □

Theorem 5.1 can be proved by applying the Nyquist stability criterion. Details of the proof can be found in [3]. Theorem 5.1 shows that an integral action, $\lambda = ((\epsilon I)/s)(L(w)z + \ell_w(w))$, to determine the price can stabilize the resulting closed-loop system. It implies that the integral plus the phase-lead compensation

$$\lambda = \frac{\epsilon}{s}\frac{\omega_L}{\omega_\ell}\frac{s+\omega_\ell}{s+\omega_L} I(L(w)z + \ell_w(w)), \qquad \omega_L > \omega_\ell > 0,$$

may be useful and stabilize the closed-loop system. By utilizing the phase-lead action, it may allow the use of a larger $\epsilon > 0$ and result in a quicker response of the closed-loop system.

5.3 EV Storage Management for Power Flow Balancing

We consider a community consisting of n agents, G_i, $i \in N = \{1, \dots, n\}$, which are households equipped with a storage charging and discharging facility. Each agent will determine the charging or discharging set-point r_i to own storage in EV. The utility that will determine the price p_i can be recognized as a manager of this community. Its priority is on the supply/demand balance and aspires to make the sum of r_i, $i \in N = \{1, \dots, n\}$, equal to the amount of supplied power w_0. We suppose that an electric power company will supply the amount w_0 based on the demand prediction. In addition to the n in-vehicle storage, we assume that the community has a stationary buffer storage and its charger/discharger G_0. The buffer storage G_0 may be necessary because of the unpredictable behavior of individual vehicles. A particular example is the sudden traveling of many vehicles. The community cannot consume sufficient power in this case, and the surplus power can be charged into the buffer storage.

Let us consider the storage dynamics

$$\dot{x}_i(t) = B_i z_i(t), \tag{5.14a}$$

$$s_i(t) = x_i(t) \qquad i \in \{0\} \cup N, \tag{5.14b}$$

where z_i is the output from PCS_i (storage charger/discharger or power conditioning system), x_i is the internal state and s_i is the state of charge which should satisfy $0 \leq s_i \leq 1$. The set-point r_i determined by G_i is the reference input to PCS_i and the

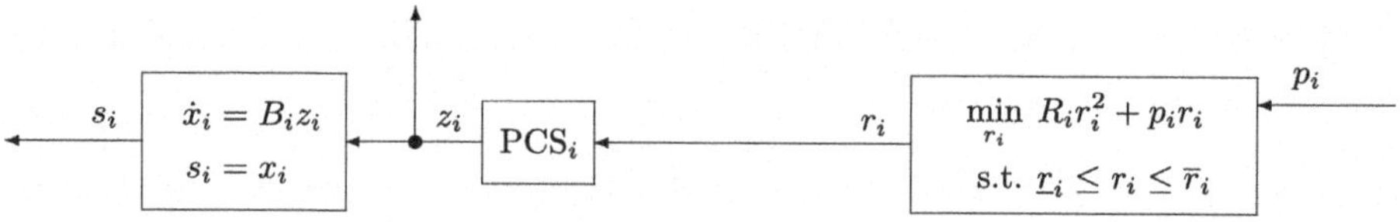

Fig. 5.2 Storage

output z_i will be charged or discharged into the storage (see Fig. 5.2). The internal current control loop of power conditioning systems usually has a very short time-constant. The time response of PCS_i is much faster than the updating period of set-point r_i, and we suppose that $z_i(t) = r_i(t)$, i.e., the dynamics of power conditioning system is $\mathsf{PCS}_i = 1$.

As for the individual optimization problem of each G_i, $i \in \{0\} \cup N$, which corresponds to (5.2), we consider

$$\min_{r_i} \quad R_i r_i^2 \tag{5.15a}$$

$$\text{subject to} \quad -r_i^{\mathrm{c}} \le r_i \le r_i^{\mathrm{c}}, \tag{5.15b}$$

where $R_i > 0$ is the weighting coefficient and the constant $r_i^{\mathrm{c}} > 0$ represents physical limitation of charging or discharging. The optimization problem in (5.4) including the price p_i can be written as

$$\min_{r_i} \quad R_i r_i^2 + p_i r_i \tag{5.16a}$$

$$\text{subject to} \quad -r_i^{\mathrm{c}} \le r_i \le r_i^{\mathrm{c}}. \tag{5.16b}$$

We note here that, as we will see in (5.20), we actually have $p_i = p_j$ for all i, $j \in \{0\} \cup N$ in the case of power flow balancing problem, since $L_i(w)$ in (5.3c) becomes 1 for all $i \in \{0\} \cup N$.

It is immediate that the optimal solution to (5.16) satisfies $r_i > 0$ if $p_i < 0$ and $r_i \le 0$ if $p_i \ge 0$. This means that, by receiving the price p_i, each agent G_i can recognize whether the utility requires charging ($r_i > 0$) or discharging ($r_i \le 0$). In the case of charging, we would like to charge the power into the in-vehicle storage and only the surplus power may charge into the buffer storage. This implies that $r_i \gg r_0$ are desired solutions. On the contrary, in the case of discharging, we would like to realize $r_i \ll r_0$. We suppose that each agent G_i will choose (switch) the weighting coefficient R_i in (5.16a) according to the sign of the received price p_i and try to realize the desired solutions

$$\begin{cases} R_i(p_i) = R_{\ell i} \ll R_{L0} = R_0(p_0), & \text{if } p_i < 0, \\ R_i(p_i) = R_{Li} \gg R_{\ell 0} = R_0(p_0), & \text{if } p_i \ge 0, \end{cases} \tag{5.17}$$

where R_{Li}, $R_{\ell i}$, R_{L0}, and $R_{\ell 0}$ are pre-specified constants.

Let us consider the constants $\underline{s}_i$ and $\bar{s}_i$ which satisfy $0 < \underline{s}_i < \bar{s}_i < 1$. The constants $\underline{s}_i$ and $\bar{s}_i$ represent the upper and lower limits of the state of charge during the operation. If the state of charge is too low, $s_i < \underline{s}_i$, the agent G_i can only accept the charging and need to consider the inequality constraint $0 \leq r_i < r_i^{\mathrm{c}}$ instead of (5.16b). On the other hand, the agent needs to consider $-r_i^{\mathrm{c}} \leq r_i \leq 0$ in the case of $s_i > \bar{s}_i$. We suppose that, by using the own state of charge s_i, each agent sets

$$\underline{r}_i(s_i) = \begin{cases} 0 & \text{if } s_i \leq \underline{s}_i, \\ -r_i^{\mathrm{c}} & \text{if } s_i > \underline{s}_i, \end{cases} \qquad \bar{r}_i(s_i) = \begin{cases} r_i^{\mathrm{c}} & \text{if } s_i < \bar{s}_i, \\ 0 & \text{if } s_i \geq \bar{s}_i, \end{cases}$$

and considers the inequality constraints $\underline{r}_i(s_i) \leq r_i \leq \bar{r}_i(s_i)$ instead of (5.16b).

The individual optimization problem of each agent G_i, $i \in N \cup \{0\}$ can be summarized as

$$\min_{r_i} \quad R_i(w_i^p)r_i^2 + p_i r_i \tag{5.18a}$$

$$\text{subject to} \quad \underline{r}_i(w_i^s) \leq r_i \leq \bar{r}_i(w_i^s), \tag{5.18b}$$

where we set $w_i = [w_i^p \quad w_i^s]^{\mathrm{T}} = [p_i \quad s_i]^{\mathrm{T}}$. By using the received price $p_i(t)$, each agent determines $r_i(t) = r_i^\flat(p_i(t))$ as a solution to (5.18).

The centralized optimization problem which corresponds to (5.3) becomes

$$\min_{\substack{r_i \\ i\in\{0\}\cup N}} \quad \sum_{i\in\{0\}\cup N} R_i(w_i^p)r_i^2 \tag{5.19a}$$

$$\text{subject to} \quad \underline{r}_i(w_i^s) \leq r_i \leq \bar{r}_i(w_i^s), \qquad i \in \{0\} \cup N, \tag{5.19b}$$

$$\mathbb{1}^{\mathrm{T}} r - w_0 = 0, \tag{5.19c}$$

where we set $\mathbb{1} = [1 \quad 1 \quad \cdots \quad 1]$ and $r = [r_i]_{i\in\{0\}\cup N}$, and w_0 denotes the amount of supplied power based on the demand prediction.

Because of the power flow balancing constraint in (5.19c), the real-time pricing strategy in (5.8) becomes

$$\dot{\lambda}(t) = \epsilon(\mathbb{1}^{\mathrm{T}} z(t) - w_0(t)), \qquad \epsilon > 0, \tag{5.20a}$$

$$p_i(t) = \lambda(t), \qquad i \in \{0\} \cup N. \tag{5.20b}$$

In the real-time pricing strategy of (5.20), the utility only needs to know the net imbalance $\mathbb{1}^{\mathrm{T}} z(t) - w_0(t)$ and each $z_i(t)$ does not necessarily need to be known. In addition, the pricing strategy does not depend on the number of vehicles that are temporarily available. This is useful in the vehicle charging/discharging management problem, since the number of available vehicles is usually unknown.

Let us consider the case $w_0 < 0$ and suppose that the buffer storage has sufficient spare capacity, i.e., $\underline{s}_0 < s_0$. The utility will require discharging to every storage, but the desired solution may be $r_0 = w_0$ and $r_i = 0, i \in N$. Even if one considers the weighting coefficients in (5.17), there may only be $-1 \ll r_i(t) < 0$ as a solution to (5.18). If one wants to obtain $r_i(t) = 0$ exactly, it may require to introduce a dead-zone nonlinearity as

$$z_i = \mathrm{PCS}_i(r_i) = \begin{cases} r_i, & \text{if } |r_i| \geq d_i, \\ 0, & \text{if } |r_i| < d_i, \end{cases} \qquad i \in N \tag{5.21}$$

where $0 < d_i \ll 1$.

The bandwidth of the closed-loop system in Fig. 5.1 is essentially determined by the step-size parameter $\epsilon > 0$ in (5.20) and the gradient $E_i(w)$ of the individual optimization problem (5.18)

$$E_i(w) = \begin{cases} -1/(2R_i(w_i^p)), & \text{if } \underline{r}_i(w_i^s) < r_i^\flat(p_i) < \bar{r}_i(w_i^s), \\ 0, & \text{if } r_i^\flat(p_i) = \underline{r}_i(w_i^s) \text{ or } r_i^\flat(p_i) = \bar{r}_i(w_i^s). \end{cases}$$

This implies that the time-constant of the closed-loop system may vary according to the operating condition of each agent (whether the current solution $r_i^\flat(p_i(t))$ is an interior point or not). If one allows additional communication between the agent G_i and utility to tell the current operating condition, the utility can have a constant bandwidth of the closed-loop system by adjusting the step-size parameter $\epsilon > 0$. An example of adjustment of $\epsilon > 0$ could be

$$\epsilon \times \sum_{i \in \{0\} \cup N} |E_i(w)| = \frac{1}{\gamma} \frac{\pi}{t_s}, \tag{5.22}$$

where t_s is the sampling period for the discretized implementation and γ is a scaling factor of the Nyquist frequency π / t_s.

5.4 Numerical Case Studies

To implement the proposed real-time pricing and decentralized decision-making methodology, we consider the discretized model of (5.20) with the sampling period $t_s = 1$ s. Each agent will update $r_i(t) = r_i^\flat(p_i(t))$ at every t_s. In the following numerical case studies, optional capabilities in (5.21) and (5.22) are also considered.

5.4.1 Numerical Case Study 1: Charging and Discharging

This numerical experiment considers a community consisting of four vehicles G_i, $i \in N = \{1, 2, 3, 4\}$, and a stationary buffer storage G_0. We use the parameters in Table 5.1 for this numerical case study. Figure 5.3 shows the resulting time response of the closed-loop system in Fig. 5.1.

Figure 5.3e shows the supplied power flow w_0. Figure 5.3a shows the output z_i from each PCS$_i$ (see (5.21)) that should track r_i determined by each G_i in a decentralized manner according to (5.18). Figure 5.3b shows the state of charge s_i of each $G_i, i \in \{0\} \cup N$. Around $0.5\,\text{h} \leq t \leq 4.0\,\text{h}$, the community has a discharging request ($w_0(t) < 0$). We can see in Fig. 5.3a that the buffer storage G_0 supplies the required power z_0 and the vehicles do not reduce their state of charge s_i, $i \in N$ (see Fig. 5.3b). However, around $t = 3.0\,\text{h}$, $s_0(t)$ reaches its lower bound $\underline{s}_0$, and G_0 stops discharging. The four vehicles immediately start discharging and share the necessary power equally. After $t = 4.0\,\text{h}$, the community has a charging request ($w_0(t) > 0$), and four vehicles start charging, while the buffer storage maintains its level. Around $t = 5.5\,\text{h}$, $s_1(t)$ reaches its upper bound $\bar{s}_1$ and G_1 stops charging. The vehicles G_i, $i = 2, 3, 4$, gradually increase their z_i and only the surplus power is charged into the buffer storage G_0. Figure 5.3c and d show the pricing signals p_i, which are determined in real-time according to (5.20), and the net imbalance $\mathbb{1}z(t) - w_0(t)$, respectively. Figure 5.3d illustrates that the supply/demand imbalance $\mathbb{1}z(t) - w_0(t) \neq 0$ at each moment of the change of operating condition smoothly recovers to 0. This confirms the effectiveness of the proposed real-time pricing and decentralized decision-making methodology.

5.4.2 Numerical Case Study 2: Plug-and-Play Type Operation

This numerical case study considers unpredictable traveling of the vehicles. We consider a community that consists of five vehicles G_i, $i \in N = \{1, 2, 3, 4, 5\}$, and a stationary buffer storage G_0. We suppose that the vehicle G_4 suddenly leaves the community at $t = 3.5\,\text{h}$, while the vehicle G_5 connects to the community at

Table 5.1 Parameters for numerical case study 1

	G_0	G_1	G_2	G_3	G_4
B_i [kWh]	400	40	40	40	40
$R_{\ell i}$	1	1	1	1	1
R_{Li}	100	100	100	100	100
r_i^c [kW]	18.0	6.0	6.0	6.0	6.0
$\bar{s}_i$	0.9	0.9	0.9	0.9	0.9
$\underline{s}_i$	0.1	0.1	0.1	0.1	0.1
$s_i(0) = x_i(0)$	0.2	0.8	0.75	0.70	0.40
d_i	–	0.2	0.2	0.2	0.2

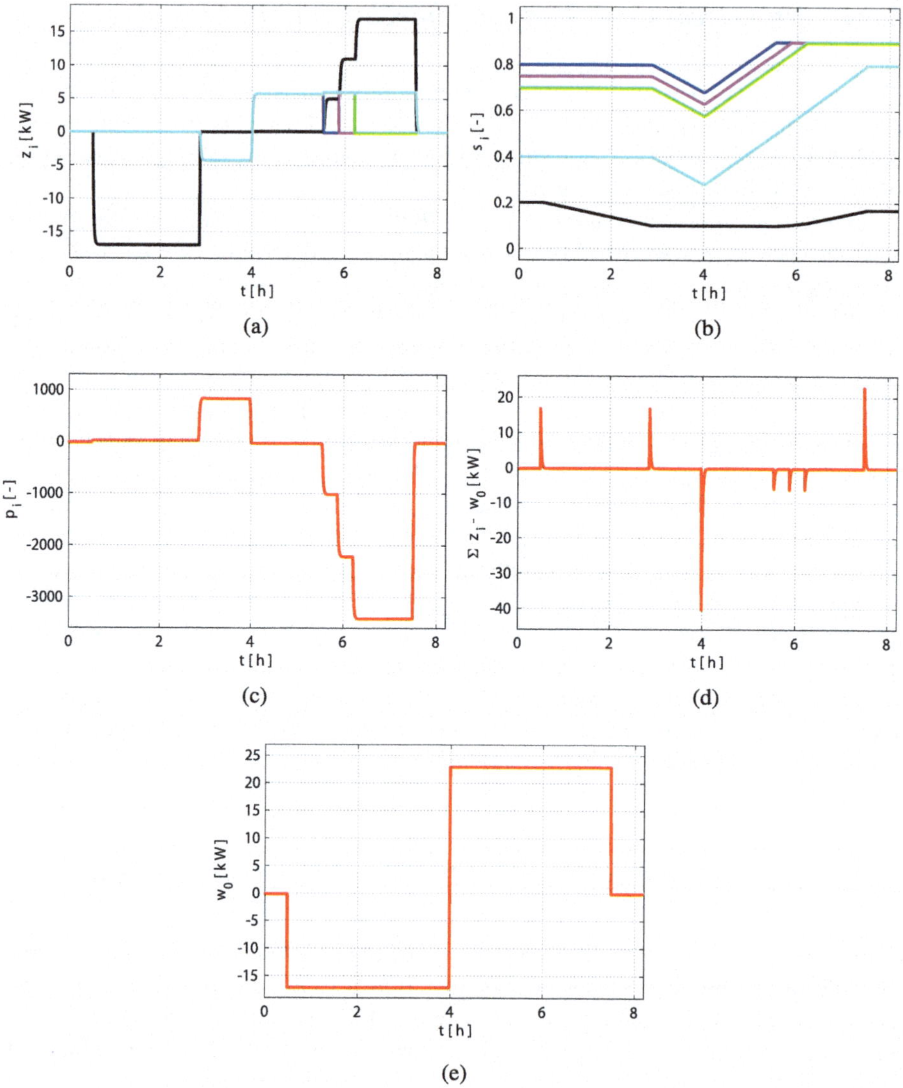

Fig. 5.3 Real-time pricing and decentralized storage charge/discharge management. (**a**) Charge/discharge flow, z_i. (**b**) State of charge, z_i. (**c**) Price, p_i. (**d**) Net imbalance, $\sum z_i - w_0$. (**e**) Supplied power flow, w_0

$t = 5.0\,\text{h}$. We use the parameters in Table 5.1, while the parameters of G_5 are the same as those of the other vehicles, and $s_5(5.0\,\text{h}) = 0.2$. We consider the supplied power flow w_0 in Fig. 5.3e. Figure 5.4 shows the resulting time response of the closed-loop system in Fig. 5.1.

Similar to the numerical case study 1, the four vehicles G_i, $i = 1, 2, 3, 4$, start discharging around $t = 3.0\,\text{h}$. At $t = 3.5\,\text{h}$, G_4 starts traveling and leaves the community. As observable, in Fig. 5.4a, that the other three vehicles increase the amount of discharging and attempt to maintain the supply/demand balance. After $t = 4.0\,\text{h}$, these three vehicles charge the power, and only the surplus power is charged into G_0. When the vehicle G_5 connects to the community at $t = 5.0\,\text{h}$, the buffer storage G_0 immediately reduces the amount of charging and the supplied power is equally shared among G_i, $i = 1, 2, 3, 5$. It confirms that the proposed real-time pricing and decentralized decision-making methodology can handle the unpredictable traveling of the vehicles without any modification of the real-time pricing strategy in (5.20).

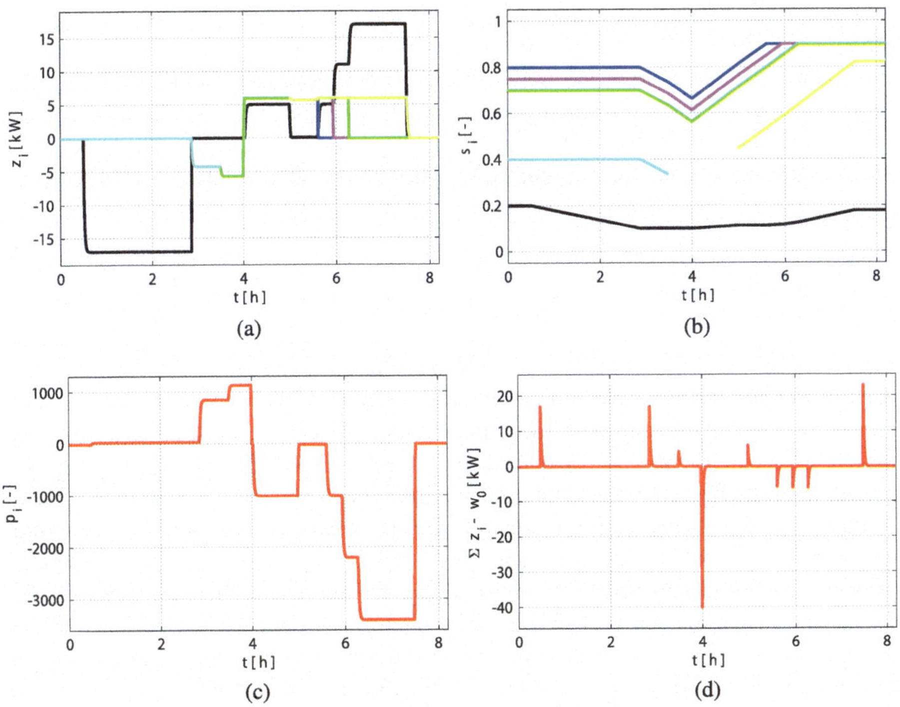

(a) (b) (c) (d)

Fig. 5.4 Real-time pricing and decentralized storage charge/discharge management with plug-and-play operation. **(a)** Charge/discharge flow, z_i. **(b)** State of charge, z_i. **(c)** Price, p_i. **(d)** Net imbalance, $\sum z_i - w_0$

Table 5.2 Parameters for numerical case study 3

	G_0	$G_i, i \in N = \{1, 2, \dots, 50\}$
B_i [kWh]	2000	40
$R_{\ell i}$	1	1
R_{Li}	1000	1000
r_i^c [kW]	180.0	6.0
$\bar{s}_i$	0.9	0.9
$\underline{s}_i$	0.1	0.1
$s_i(0) = x_i(0)$	0.3	$0.15 + 0.15i$
d_i	–	0.2

5.4.3 Numerical Case Study 3: Community Consists of 50 Vehicles

This numerical case study considers a community consisting of 50 vehicles and a stationary buffer storage. The parameters for this numerical case study can be found in Table 5.2. The resulting time response of the closed-loop system in Fig. 5.1 is illustrated in Fig. 5.5.

Figure 5.5b and f show the scale-up of Fig. 5.5a and e, respectively. Since the decision-making of each vehicle is completely decentralized, the proposed real-time pricing and decentralized decision-making methodology was implementable without any increase of the computational burden, even considering 50 vehicles.

5.5 Conclusions

Motivated by the decentralized charging/discharging management for EV storage, this chapter investigated the optimal regulation problem with steady-state constraints under decentralized decision-making. Each agent (EV) is allowed to determine its own charging or discharging set-point and the utility, which corresponds to a manager of the community, attempts to realize a socially optimized solution that fulfills the supply/demand power flow balance. To align the individual decision-making of each agent with the socially optimal solution, the utility provides an additional price, which conceptually represents a tax or subsidy for the agent. The proposed real-time pricing strategy is heuristic, since it combines the gradient method to determine the price, individual optimization, and dynamics of the agents. However, we have shown that the resulting closed-loop system is stable at least locally. The numerical case studies illustrated the effectiveness of the proposed real-time pricing and decentralized decision-making methodology. The other applications of the real-time pricing and decentralized decision-making methodology include the optimal power flow problem of power grids [4] and the voltage deviation suppression for PV generation plants [5].

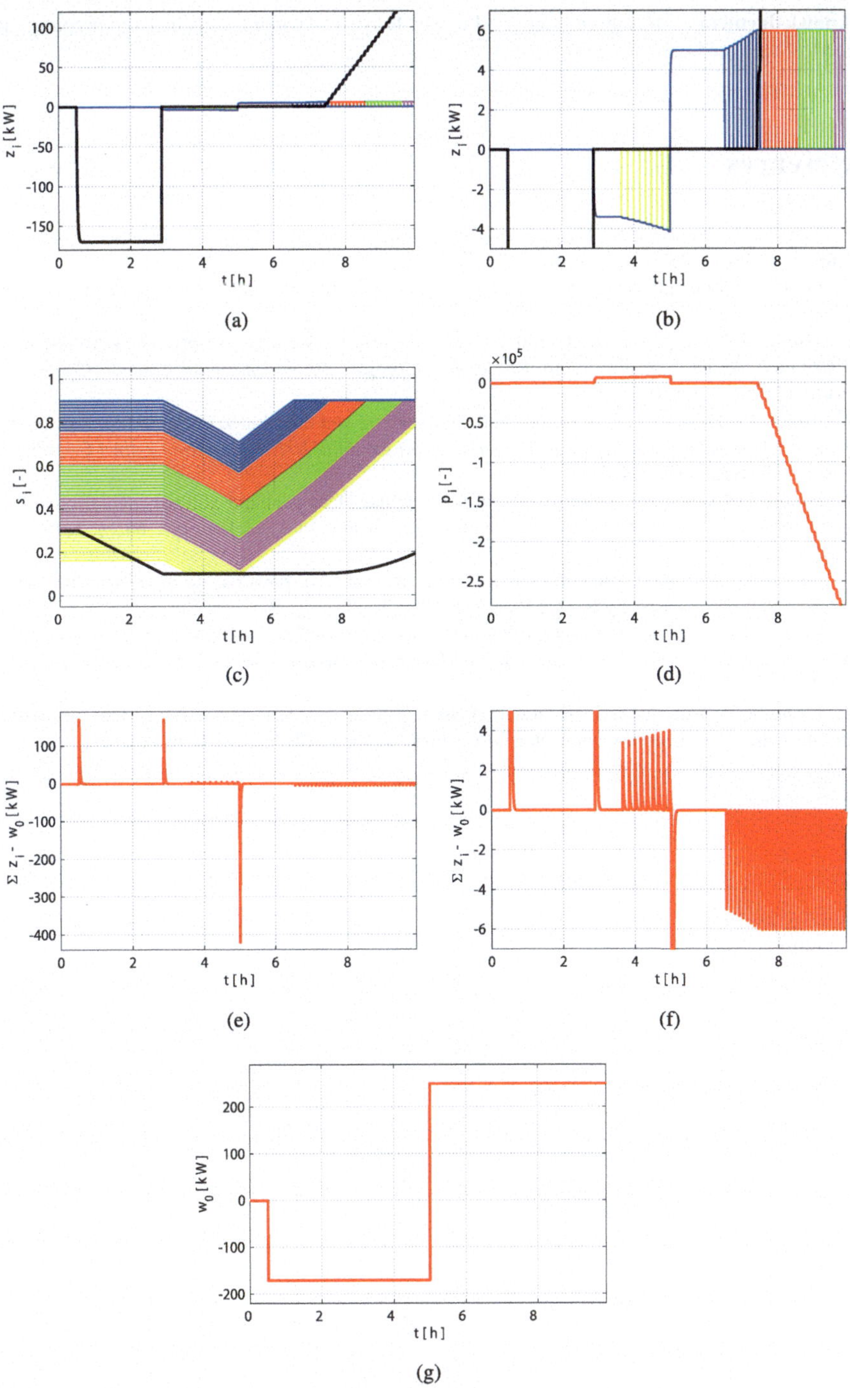

Fig. 5.5 Real-time pricing and decentralized storage charge/discharge management for the community consists of 50 vehicles. (**a**) Charge/discharge flow, z_i. (**b**) Charge/discharge flow, z_i. (**c**) State of charge, z_i. (**d**) Price, p_i. (**e**) Net imbalance, $\sum z_i - w_0$. (**f**) Net imbalance, $\sum z_i - w_0$. (**g**) Supplied power flow, w_0

Acknowledgements The author is grateful to Mr. Junya Terayama for his helps in preparing the numerical case studies.

References

1. K.J. Arrow, L. Hurwicz, H. Uzawa, *Studies in Linear and Non-linear Programming* (Stanford University Press, Palo Alto, 1958)
2. S. Boyd, L. Vandenberghe, *Convex Optimization* (Cambridge University Press, Cambridge, 2004)
3. K. Hirata, J.P. Hespanha, K. Uchida, Real-time pricing leading to optimal operation under distributed decision makings, in *Proceedings of the 2014 American Control Conference*, Portland (2014), pp. 1952–1932
4. K. Hirata, J.P. Hespanha, K. Uchida, Real-time pricing and distributed decision makings leading to optimal power flow of power grids, in *Proceedings of the 2015 American Control Conference*, Chicago (2015), pp. 2284–2291
5. K. Hirata, H. Akutsu, A. Ohori, N. Hattori, Y. Ohta, Decentralized voltage regulation for PV generation plants using real-time pricing strategy. IEEE Trans. Ind. Electron. **64**(6), 5222–5232 (2017)
6. A. Jokić, M. Lazar, P.P.J. van den Bosch, Real-time control of power systems using nodal prices. Int. J. Electr. Power Energy Syst. **31**(9), 522–530 (2009)
7. N. Li, L. Chen, S.H. Low, Optimal demand response based on utility maximization in power networks, in *Proceedings of the 2011 IEEE Power and Energy Society General Meeting* (2011), pp. 1–8
8. K. Tsumura, H. Yamamoto, D. Kiribuchi, Generalization of Uzawa algorithm and its application, in *The 13th SICE Annual Conference on Control Systems* (2013), 8D3-6, in Japanese

Chapter 6
A Scalable Control Approach for Providing Regulation Services with Grid-Integrated Electric Vehicles

Stijn Vandael, Tom Holvoet, Geert Deconinck, Hikari Nakano, and Willett Kempton

6.1 Introduction

Growing concerns about the environment and increasing fuel prices are causing a shift toward electric vehicles (EVs). Because EVs are idle most of the day, they offer a lot of flexibility in their charging timing and rate. EV aggregators are typically seen as actors using this charging flexibility [1].

A wide range of opportunities exist for intelligently controlling EV charging with an EV aggregator. Examples are day-ahead load scheduling [2] and providing ancillary services [3, 4].

In this chapter, we focus on EV aggregators that provide regulation as an ancillary service for a transmission system operator (TSO) [5]. To operate an EV fleet as a regulation reserve, an aggregator needs to respond to a TSO's regulation-up signal (e.g., charge less or inject more) or regulation-down signal (e.g., charge more or inject less) within seconds. Therefore, an aggregator controls power transfer for each

S. Vandael
Lead software architect, Nuvve Corporation, San Diego, CA, USA

T. Holvoet
Department of Computer Science (CW), KU Leuven, Belgium

G. Deconinck
Department of Electrical Engineering (ESAT), KU Leuven, Belgium

H. Nakano (✉)
Algorithm Engineer, Nuvve Corporation, San Diego, CA, USA
e-mail: hikari@nuvve.com

W. Kempton
College of Earth, Ocean, and Environment, University of Delaware, Newark, DE, USA
e-mail: willett@udel.edu

T. Suzuki et al. (eds.), *Design and Analysis of Distributed Energy Management Systems*, Power Electronics and Power Systems,
https://doi.org/10.1007/978-3-030-33672-1_6

EV using a GIV control approach. In this paper, a GIV control approach is defined as the set of algorithms, models, and interactions an aggregator uses to control bidirectional power transfer between EVs and the electricity grid. In the current literature, several GIV control approaches have been proposed for controlling EVs as a regulation reserve [4]. These control approaches typically calculate a unique charging schedule for each EV; however, as the number of EVs increases, such conventional approaches will face increasing scheduling complexity. This complexity is particularly apparent because of the scheduling problem's dynamic nature, which is characterized by EV owners plugging in and unplugging EVs at varying times.

To cope with these challenges, we propose a GIV control approach that calculates charging schedules per-bin rather than per-vehicle. Each bin contains hundreds of EVs with similar charging characteristics (e.g., departure time, energy-to-be-charged, maximum power). The key contributions of this paper are:

1. Problem formulation of controlling many EVs as a regulation capacity (Sect. 6.3).
2. Describing a GIV control approach that achieves scalability by gathering EVs in bins (Sect. 6.4).
3. Evaluating the scalability and scheduling quality of the GIV control approach through benchmarking against fully centralized and fully distributed scheduling approaches (Sect. 6.5).

6.2 Related Work

The related work of this paper is divided in two parts. In the first part, we give an overview of papers that describe algorithms to provide regulation services. In the second part, we provide an overview of papers describing heuristic scheduling approaches for demand response (DR).

6.2.1 Regulation Services

In most work concerning regulation services provision, an individual device model is used to calculate a plan. Our approach, which aggregates EVs with similar charging characteristics in bins, uses these per-EV scheduling approaches as a benchmark.

Sortomme et al. [4, 6] formulated the problem of scheduling grid-integrated EVs in the ERCOT region as a linear program (LP). In this formulation, EVs are scheduled to provide regulation services, spinning reserve services, and charging at hourly price tariffs. These objectives are all gathered in a single objective function to maximize aggregator profit. Simulation results demonstrate that the LP approach offers significant financial benefits to customers and aggregators.

In our work, a per-EV optimization is used as a benchmark for our bin-based GIV approach. Furthermore, the approach uses mixed-integer quadratic programming, which combines a quadratic objective for peak load reduction with integer variables for bids with a fixed regulation granularity (Sect. 6.3). In contrast to the ERCOT region, variable-sized bids are not allowed in the PJM regulation market.

Al-Awami and Sortomme [7] proposed a mixed-integer stochastic linear programming method for scheduling grid-integrated EVs. In this work, EVs were scheduled for trading in a pool-based energy market while providing regulation services. The overall objective function was to maximize aggregator profit and maintain trading risk by using conditional value at risk. Simulation results showed that expected profits can be increased while lowering trading risk. In our problem description, a per-EV optimization is used as a benchmark. Nonetheless, no stochastic variables (e.g., market uncertain-ties or stochastic generation) are included in our problem formulation as this would demand even more computational resources, which highlights the importance of the identified challenges for large-scale control (Sect. 6.3.3).

6.2.2 Heuristic Scheduling for DR

In this paper, we use bins as a heuristic scheduling approach to provide regulation services for many EVs. An important challenge in heuristic scheduling approaches for DR is calculating near-optimal schedules for many devices while limiting resource usage (e.g., computational and communication resources). This section provides a representative selection of heuristic scheduling approaches for DR.

Koch et al. [8] proposed a bin-based approach for controlling many heterogeneous thermostatically controlled loads (TCLs) such as refrigerators and electric water heaters. To simulate the behavior of thousands of TCLs, a bin transition modeling technique that approximates the aggregate dynamic behavior of the TCL population was used. To control TCLs, this model was used in an model predictive control framework. Results showed that the model is accurate enough to let a population of 1000 TCLs track a pre-defined signal. In this work, bins were defined by TCL temperature ranges, transitions between bins were non-deterministic (probability mass), and control signals were defined by on/off switching probabilities. Although this approach differs significantly from a bin-based approach for EVs, we also aim to improve scalability by classifying devices in bins.

6.3 Problem Formulation

The main stakeholder in our problem formulation is an aggregator that controls the bidirectional power transfer of contracted EVs. The aggregator's goal is to provide regulation services while complying with driver energy requirements.

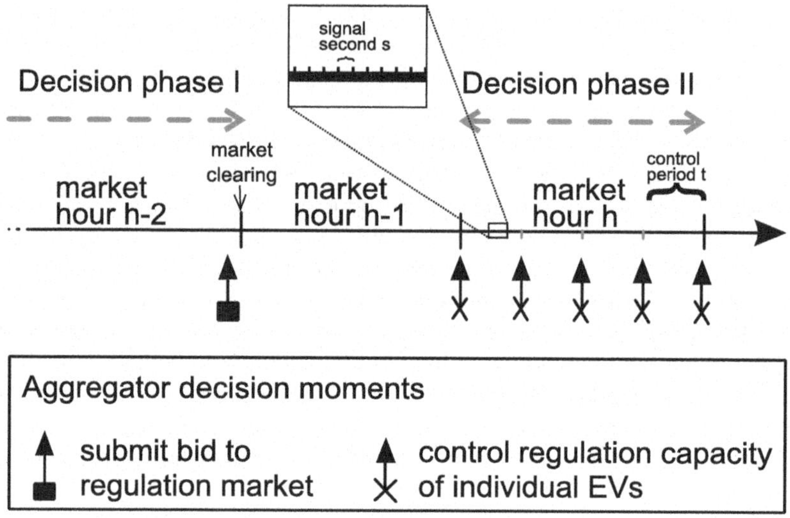

Fig. 6.1 Decision making phases of the aggregator

To achieve this goal, the aggregator makes decisions in two phases (Fig. 6.1). In the first decision phase (hour h-2), the aggregator submits a bid to the regulation market. During the second decision phase, the aggregator controls individual EVs to comply with the bid submitted in the first decision phase. Although regulation market rules slightly differ between TSOs [9], the problem formulation in this paper is based on the market rules of PJM, the world's largest TSO [10].

6.3.1 Decision Phase I

In the first decision phase, the aggregator decides how much regulation capacity P_h^{bid} to bid for market period h. The unit of P_h^{bid} is kWh, which is the ability to provide 1 kW of symmetrical regulation power (equal regulation-up and regulation-down) for 1 h. A bid for market hour h is defined as:

$$P_h^{\text{bid}} = n_h^{\text{bid}} \, \Delta P_{\text{bid}}, \qquad n_h^{\text{bid}} \in \mathbb{N}_0, \tag{6.1}$$

where n_h^{bid} is the bid size and ΔP_{bid} is the bid granularity, which is a regulation market property. A bid for market hour h must be submitted before the end of hour $h - 2$, which is when the regulation market clears for hour h. We assume that a bid will always be accepted, which is a reasonable assumption as an EV aggregator can offer significantly lower prices than conventional regulation reserves (e.g., gas turbines) due to low operating costs.

An aggregator receives a payment from the TSO for P_h^{bid}, based on the market clearing price for hour h. Clearing prices in the regulation market unpredictably fluctuate [11]; therefore, we assume that no clearing price predictions are available. Instead, an aggregator tries to maximize the regulation capacity P_h^{bid} in its bids independent of any pricing scheme.

After decision phase I, the aggregator must then provide the promised regulation capacity in decision phase II.

6.3.2 Decision Phase II

During decision phase II (market period h), the aggregator communicates with its EVs to control the fleet's total regulation capacity. Controlling the online regulation capacity of an EV fleet typically happens on a shorter time scale than market orders; therefore, each market period h is divided into equally spaced control periods (Fig. 6.1). In 1 day, there are a total of T control periods. To comply with a regulation bid, the fleet's regulation capacity should be at least as much as contained within the bid:

$$P_s^{\text{reg}} \geq P_h^{\text{bid}} \mid h = \mathbf{H}(s), \qquad \forall s \in \{1, \ldots, S\}, \tag{6.2}$$

where the function $\mathbf{H}$ maps a signal periods s to its market period h.

The fleet's total regulation capacity P_s^{reg} is the sum of the regulation capacities of each EV. In this work, we assume individual EVs only advertise a symmetric regulation capacity (equal regulation-up and regulation-down amounts) because a symmetric regulation bid is required:

$$P_s^{\text{reg}} = \sum_{i=1}^{N_{\text{ev}}} P_{i,s}^{\text{reg}}, \qquad \forall t \in \{1, \ldots, S\}, \tag{6.3}$$

wherein EVs not connected to the grid provide no regulation capacity.

The regulation capacity provided by a single EV is determined by its preferred operating point (POP). Three examples of different POPs and matching regulation capacities are shown in Fig. 6.2. POP sign conventions are based on the regulation capacity provided by power generators: a positive POP is discharging (example 1) and a negative POP is charging (example 3). Maximum regulation capacity is

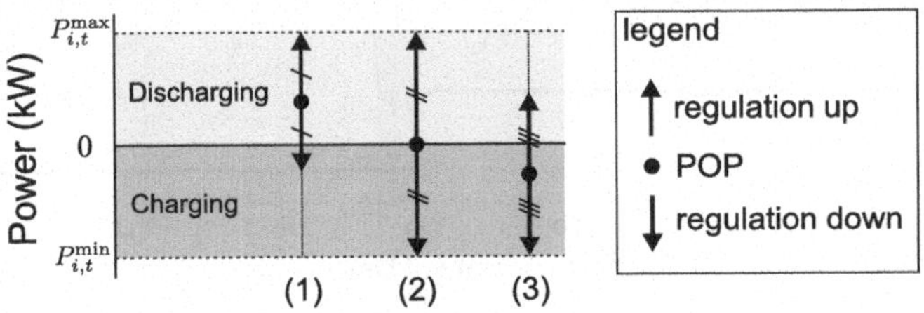

Fig. 6.2 Examples of symmetric regulation capacity of an EV. POP > 0 (1), POP = 0 (2) and POP < 0 (3)

obtained at a POP = 0 (example 2). As can be seen from these examples, for EV i at time s, the POP unambiguously defines the symmetric regulation capacity $P_{i,s}^{\text{reg}}$, given minimum charging power P_i^{min} and maximum discharging power P_i^{max}:

$$P_{i,s}^{\text{reg}} = \min(P_i^{\text{max}} - P_{i,s}^{\text{pop}},\ P_{i,s}^{\text{pop}} - P_i^{\text{min}}), \quad \forall i \in \{1, \ldots, N_{\text{ev}}\}. \tag{6.4}$$

Consequently, total regulation capacity can be adjusted by controlling the POP of each EV in a fleet. However, a challenge for an aggregator is controlling each EV's POP to provide sufficient regulation capacity while still complying with EV users' energy requirements. To avoid burdening EV batteries with extra charge cycles, which reduce battery life, POP is only allowed to be 0 or positive($P_{i,s}^{\text{pop}} \geq 0$).

Although an EV's regulation capacity defines the potential to adjust its charging power around a chosen POP, a separate regulation signal P_s^{sig} determines how much of the regulation capacity must actually be deployed during signal period s. Therefore, each control period t is divided into equally spaced signal periods (Fig. 6.1). In this paper, the signal period duration is 1 s because typical regulation signals change at a rate of ~2–10 s [10].

An example of a regulation signal P_s^{sig} from PJM resulting from two market bids (100 and 200 kWh) is shown in Fig. 6.3. This signal is known to be approximately energy neutral, meaning that PJM is expected to request an equal amount of up-regulation and down-regulation during one market period.

The aggregator follows a regulation signal using two steps (Fig. 6.4). In the first step, the aggregator sums individual regulation capacities as in (6.3). In the second step, the aggregator calculates the fraction μ_s of the available total regulation capacity P_s^{reg}, that is deployed to follow the regulation signal P_s^{sig} in signal period s:

$$\mu_s = \frac{P_s^{\text{sig}}}{P_s^{\text{reg}}}. \tag{6.5}$$

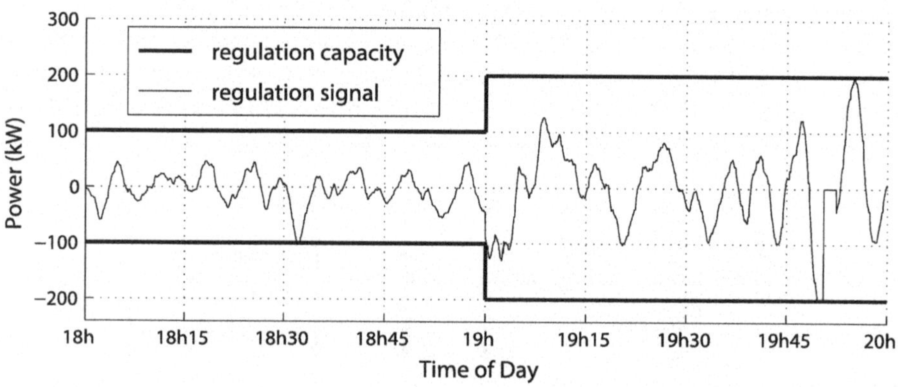

Fig. 6.3 Example of two regulation bids (100 and 200 kWh) and PJM's resulting regulation signal

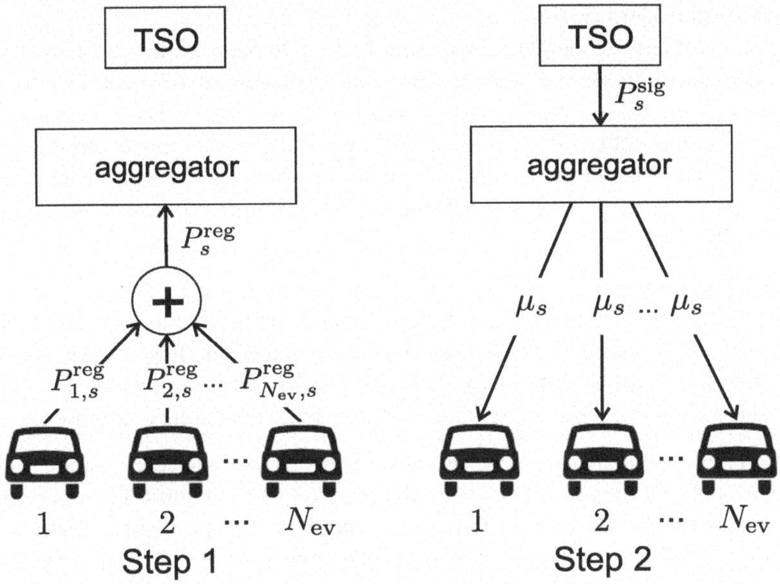

Fig. 6.4 The aggregator sums all individual regulation capacities to obtain the fleet's total regulation capacity (step 1) and linearly divides the received regulation signal among EVs, based on the total regulation capacity (step 2)

When EVs receive μ_s from the aggregator, each EV deploys this fraction of their individual regulation capacity. The total response to the regulation signal in regulation period s is:

$$P_s^{\mathrm{resp}} = \mu_s P_{1,s}^{\mathrm{reg}} + \mu_s P_{2,s}^{\mathrm{reg}} + \cdots + \mu_s P_{N_{\mathrm{ev}},s}^{\mathrm{reg}}. \tag{6.6}$$

Note that in case $\mu > 1$, the regulation signal oversteps the regulation capacity. In this case, the aggregator promised too much regulation capacity in its bids, which violates constraint (6.2). This will cause a poor quality of response, which is quantified by the TSO by a performance score [10]. Consistent low performance scores will disqualify an EV fleet from providing regulation services.

In summary, in decision phase II, the aggregator decides the individual regulation capacity of each EV by controlling their POP. Based on the fleet's total available regulation capacity, a regulation signal is then divided between the EVs.

6.3.3 Challenges in a Large-Scale Setting

An aggregator's decision problem aggregator consists of bidding on the regulation market (decision phase I) and controlling its EVs' regulation capacities (decision phase II). Additional challenges for solving this problem for many EVs are twofold: the decision problem's computational complexity of (1) and high fleet dynamics (2).

Computational Complexity
A decision problem' s computational complexity determines the timespan in which a new decision can be calculated. This timespan should be kept low to allow decision-making in a timely manner. However, for many EVs, computational complexity can quickly become excessive, as each EV adds more variables to the control problem. Furthermore, this complexity is worsened through the presence of discrete variables in (6.1), resulting in a discrete optimization problem, which is typically NP-hard.

EV Fleet Dynamics
The decisions in both phases depend on the energy requirements of individual EVs in the fleet. A challenge herein is the dynamic nature of these requirements. In a large fleet, EVs are continuously being plugged in and unplugged. To correctly handle these dynamics, the aggregator must revise its decisions continuously.

Both challenges are apparent because they exacerbate each other. Greater computational complexity makes it harder to react to a dynamic fleet, and greater fleet dynamics make decision-making more complex. To cope with these challenges, we propose a GIV mechanism based on an approximate bin model that keeps computational complexity low (challenge 1) and dynamically recalculates decisions (challenge 2).

6.4 Bin-Based GIV Control Approach

In this section, we present a bin-based GIV control approach to the aggregator problem formulated in Sect. 6.3. The key advantage of this approach is its ability to control many EVs by classifying EV characteristics in bins.

The description of this GIV approach has a bottom-up structure. First, the individual EV behavior is explained, then the binning mechanism is explained, and finally the scheduling mechanism is described (Fig. 6.5). EV behavior is modeled by a Markov decision process (MDP), the binning mechanism is used by EVs to classify their POP scheduling flexibility in bins, and the optimization mechanism is used by the aggregator to calculate each bin's POP schedule.

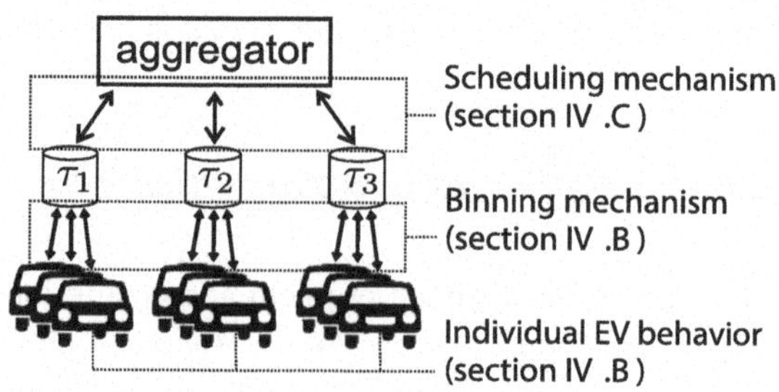

Fig. 6.5 Dual mechanism design of the GIV control approach

6.4.1 EV Charging Behavior

EV charging behavior determines how an EV follows an aggregator's requested POP while complying with driver energy requirements. The complete EV charging behavior for a single EV is visualized in a state of charge (SOC) graph (Fig. 6.6). This graph shows an EV's charging cycle, which runs from arrival to departure time. The outer boundaries in this graph, known as the grid-integrated boundaries, define valid SOC values for an EV driver. These boundaries are determined by the driver's minimum and requested energy, with the minimum energy being that needed during the entire charging cycle for an unplanned trip and the required energy being the energy required at departure time for a planned trip. For an EV fleet of size N_{ev}, the maximum and minimum grid-integrated boundaries of each EV connected to the grid can be expressed as:

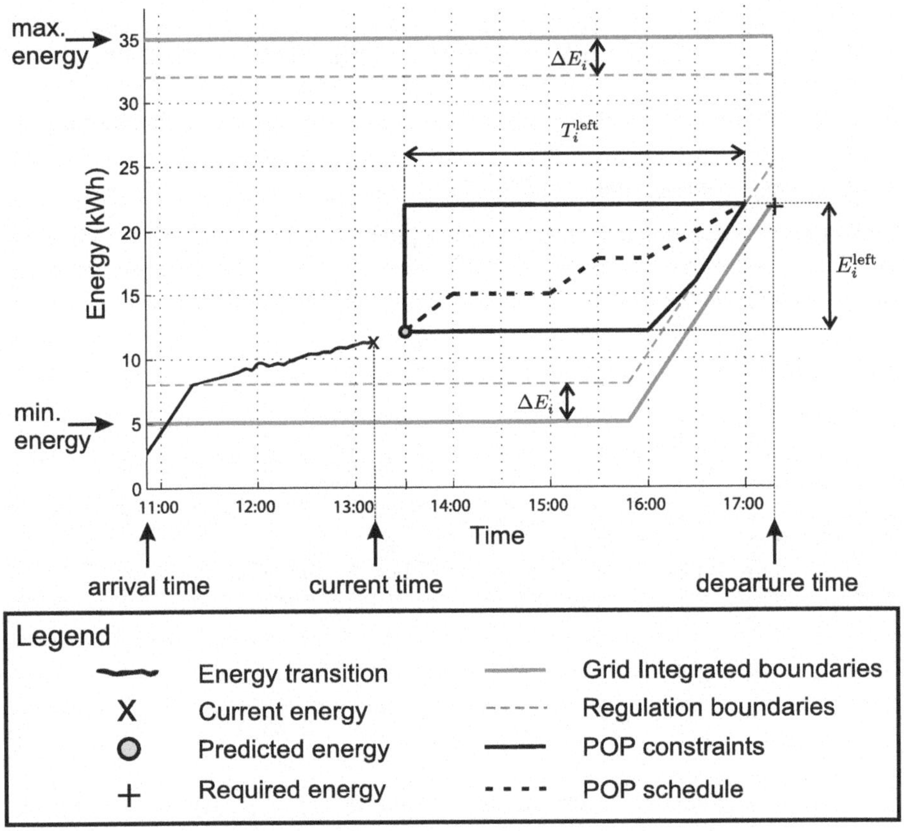

Fig. 6.6 Example of a SOC graph for an EV

$$\mathbf{E}_i^{\text{gi+}} = \left\{E_{i,s}^{\text{gi+}} \mid E_{i,s}^{\text{gi+}} = E_i^{\text{cap}}, \forall s \in \left\{S_i^{\text{arr}}, \ldots, S_i^{\text{dep}}\right\}\right\}, \tag{6.7}$$

$$\mathbf{E}_i^{\text{gi-}} = \left\{E_{i,s}^{\text{gi-}} \mid E_{i,s}^{\text{gi-}} = \max\left(E_i^{\text{req}} - P_i^{\max}\left(S_i^{\text{dep}} - s\right), E_i^{\min}\right),\right.$$
$$\left.\forall\, s \in \left\{S_i^{\text{arr}}, \ldots, S_i^{\text{dep}}\right\}\right\}, \tag{6.8}$$

$$\forall\, i \in \mathscr{F}_{c,s_{\text{cur}}},$$

where:

- S_i^{arr} is the arrival time (in seconds).
- S_i^{dep} is the departure time (in seconds).
- E_i^{req} is the absolute energy required by EV i before the departure time.
- $P_i^{\max}$ is the maximum power at which an EV i can charge its battery. $P_i^{\max}$ is dependent on the battery's power limitations, the battery management system (BMS), and the local electrical grid connection (e.g., EVs can charge at a higher power through a fast charging station than through a regular household connection).
- $E_i^{\min}$ is the minimum energy, required by the EV driver for unplanned trips.
- E_i^{cap} is the EV's battery capacity.
- $\mathscr{F}_{c,s_{\text{cur}}}$. is the set of connected EVs at time s_{cur}.

To model the complete decision-making process of an EV, each EV's charging behavior is represented as an MDP. The state space E^{soc} of an EV is composed of the SOC and is defined as:

$$E^{\text{soc}} = \left\{e \in \mathbb{R} \mid e \in \left[0, E_i^{\text{cap}}\right]\right\}, \tag{6.9}$$

with E_i^{cap} representing battery capacity. A full charging cycle of EV i starts in the initial state $\{e_s \mid e_s = E_i^{\text{init}}, s = S_i^{\text{arr}}\}$, and ends in the terminal state $\{e_s \mid E_i^{\text{req}} \leq e_s \leq E_i^{\text{cap}}, s = S_i^{\text{dep}}\}$. Action space P consists of all POP values for charging an EV, defined as $P = \{p \in \mathbb{R} \mid p \in [P_i^{\min}, P_i^{\max}]\}$.

The system dynamics of EV i are described by the state transitions $E_{i,s}^{\text{soc}} \rightarrow E_{i,s+1}^{\text{soc}}$:

$$\forall s \in [S_i^{\text{arr}}, S_i^{\text{arr}} + \Delta s, \ldots, S_i^{\text{dep}}],$$
$$E_{i,s+1}^{\text{soc}} = E_{i,s}^{\text{soc}} - P_{i,s}^{\text{pop}} \Delta s - \mu_s P_{i,s}^{\text{reg}} \Delta s \mid t = \mathbf{T}(s), \tag{6.10}$$

where the SOC changes depending on charging power $P_{i,s}^{\text{chrg}}$ and the fraction μ_s of the regulation capacity $P_{i,s}^{\text{reg}}$ deployed. The POP $P_{i,s}^{\text{pop}}$ of EV i at second s is determined by policy π:

$$P_{i,s}^{\text{pop}} = \pi\left(E_{i,s}^{\text{soc}}, E_{i,s}^{\text{gi-}}, P_i^{\min}, P_{i,s}^{\text{req}}\right)$$
$$= \begin{cases} P_i^{\min}, \text{ if } E_{i,s}^{\text{soc}} < E_{i,s}^{\text{gi-}}, \\ P_{i,s}^{\text{req}}, \text{ if } E_{i,s}^{\text{soc}} \geq E_{i,s}^{\text{gi-}}. \end{cases} \tag{6.11}$$

The input parameter $P_{i,s}^{\text{req}}$ to policy π is the POP requested by the binning mechanism. When the SOC is within the grid-integrated boundaries of the EV, this POP can be followed.

6.4.2 *Binning Mechanism*

The goal of the binning mechanism is to provide a POP as input to the policy in (6.11) to provide regulation capacity for the aggregator. The regulation boundaries determine an area of SOC values usable for following a regulation signal (Fig. 6.6). These boundaries are defined as an offset ΔE_i from the grid-integrated boundaries:

$$\mathbf{E}_i^{\text{reg+}} = \left\{E_{i,s}^{\text{reg+}} \mid E_{i,t}^{\text{reg+}} = E_i^{\text{gi+}} - \Delta E_i, \quad \forall\, s \in \left\{s_i^{\text{arr}}, \ldots, s_i^{\text{dep}}\right\}\right\}, \tag{6.12}$$

$$\mathbf{E}_i^{\text{reg-}} = \left\{E_{i,s}^{\text{reg-}} \mid E_{i,s}^{\text{reg-}} = E_i^{\text{gi-}} + \Delta E_i, \quad \forall\, s \in \left\{s_i^{\text{arr}}, \ldots, s_i^{\text{dep}}\right\}\right\}, \tag{6.13}$$

$$\forall\, i \in \mathscr{F}_{c,s_{\text{cur}}}.$$

Regulation boundaries are intended to guarantee a margin ΔE for following a regulation signal that is chosen based on a worst-case scenario. In this worst-case scenario, the aggregator requests a maximum regulation-down or regulation-up value for a certain period Δs_{sig}. Based on this period, ΔE_i can be determined:

$$\Delta E_i = P_i^{\max} \Delta s_{\text{sig}}. \tag{6.14}$$

In the example shown in Fig. 6.6 is $\Delta E_i = 3$ kWh, based on a maximum charging power of 12 kW and Δs_{sig} of 900 s.

If an EV arrives with an SOC outside of regulation boundaries, the POP is controlled to obtain an SOC within regulation boundaries as quickly as possible. Inside regulation boundaries, the binning mechanism uses a request-response procedure to communicate with its respective bin. First, an EV sends a request message containing its binned charging requirement (known as binned POP constraints), which approximates its charging requirements. The binning mechanism then determines the current bin's POP schedule and sends it to the EV. This POP schedule's control value is used in the EV's policy in (6.11). This is a light-weight interaction (limited calculation time and communication overhead) that allows each EV to quickly adapt its charging power to a changing environment.

An EV's actual charging at requirements at time s is defined as POP constraints that define the SOC area in which the POP can be scheduled (Fig. 6.6). A POP schedule within these constraints starts from the predicted SOC, the SOC predicted for the beginning of the upcoming control period, and ends in the required SOC, which is defined by the driver at departure time. In Fig. 6.6, the predicted SOC is 12 kWh, and the required SOC is 22 kWh.

As an EV follows a regulation signal, predicted SOC is continuously adjusted, based on the current SOC $E^{\text{soc}}_{i,s}$ and the POP $P^{\text{POP}}_{i,t}$ scheduled for the current control period t. For a time s_{cur} within a control period t:

$$E^{\text{pred}}_{i,t+1} = E^{\text{soc}}_{i,s_{\text{cur}}} - \frac{P^{\text{pop}}_{i,t}\,\Delta s_{\text{left}}}{S/24} \mid t = \mathbf{T}(s_{\text{cur}}), \tag{6.15}$$

$$\Delta s_{\text{left}} = (t+1)\,\Delta t - s_{\text{cur}}\Delta s, \tag{6.16}$$

where Δs_{left} is the number of seconds left in control period t. As $\Delta s_{\text{left}} \rightarrow 0$, the end of the control period t comes near and the SOC prediction improves.

Based on the predicted SOC, required SOC, and departure time, energy and time left to be charged (Fig. 6.6) can be calculated for the current time s_{cur}:

$$E^{\text{left}}_{i} = E^{\text{req}}_{i} - E^{\text{pred}}_{i,t} \mid t = \mathbf{T}(s_{\text{cur}}), \tag{6.17}$$

$$T^{\text{left}}_{i} = \mathbf{T}(S^{\text{dep}}_{i} - \Delta s_{\text{sig}}) - \mathbf{T}(s_{\text{cur}}) - 1. \tag{6.18}$$

Based on these values, the minimum and maximum POP constraints of each EV can be expressed as:

$$\mathbf{E}^{\text{pop+}}_{i,s_{\text{cur}}} = \left\{ E^{\text{pop+}}_{i,t} | E^{\text{pop+}}_{i,t} = E^{\text{left}}_{i,s_{\text{cur}}}, \forall\, t \in \left\{1, \ldots, T^{\text{left}}_{i,s_{\text{cur}}}\right\}\right\}, \tag{6.19}$$

$$\mathbf{E}^{\text{pop-}}_{i,s_{\text{cur}}} = \left\{ E^{\text{pop-}}_{i,t} \mid E^{\text{pop-}}_{i,t} = \max(E^{\text{left}}_{i,s_{\text{cur}}} - P^{\max}_{i}(T^{\text{left}}_{i,s_{\text{cur}}} - t), 0), \right.$$
$$\left. \forall\, t \in \left\{1, \ldots, T^{\text{left}}_{i,s_{\text{cur}}})\right\}\right\}, \tag{6.20}$$

$$\forall\, i \in \mathscr{F}_{c,s_{\text{cur}}}.$$

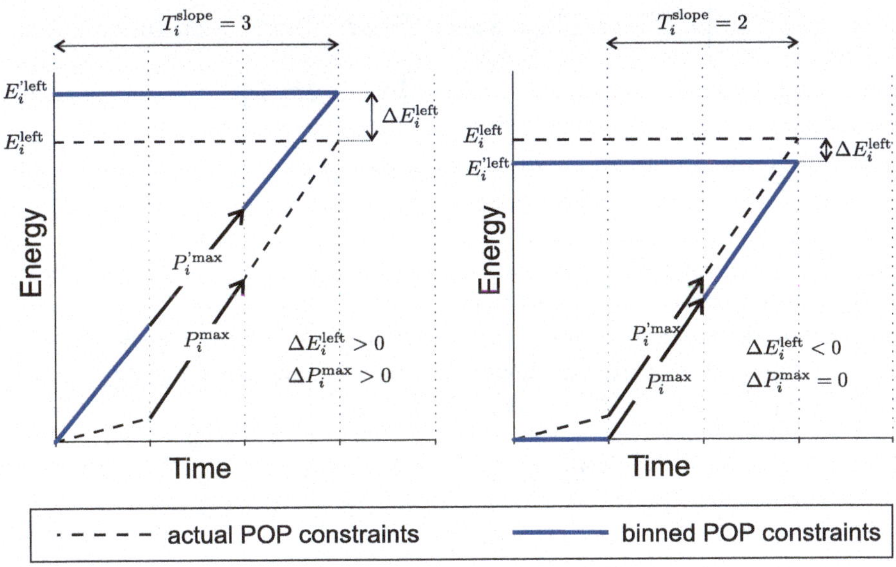

Fig. 6.7 Example of rounding up to $T_i^{\text{slope}} = 3$ and down to $T_i^{\text{slope}} = 2$

In contrary to grid-integrated and regulation boundaries, POP constraints are defined with granularity Δt, and are relative to the beginning of the next control period and predicted energy $E_{i,s_{\text{cur}}}^{\text{pred}}$.

To determine an EV's binned charging requirements, its POP constraints must first be approximated (or binned) into so-called binned POP constraints (Fig. 6.7). In a binned POP constraint, the parameters E_i^{left} and P_i^{max} in (6.19) and (6.20) are replaced with the parameters $E_i^{\prime\text{left}}$ and $P_i^{\prime\text{max}}$, which have the following property:

$$\frac{E_i^{\prime\text{left}}}{P_i^{\prime\text{max}}\Delta t} = T_i^{\text{slope}} \mid T_i^{\text{slope}} \in \mathbb{N}_0, \tag{6.21}$$

where T_i^{slope} is a whole number that represents the amount of control periods required to charge the amount of energy left to be charged $E_i^{\prime\text{left}}$ at power $P_i^{\prime\text{max}}$. If T_i^{slope} is a whole number, the POP constraint is binned. In Fig. 6.7, a non-binned POP constraint can easily be identified by the changing slope, which indicates that T^{slope} is not a whole number.

An EV determines its binned POP constraints by calculating both δ_{up} and δ_{down} and uses the minimum $\delta_{\min} = \min(|\delta_{\text{up}}|, |\delta_{\text{down}}|)$ to induce the smallest relative error. Based on this value, $E_i^{\prime\text{left}}$ and $P_i^{\prime\text{left}}$ are determined and used to construct the binned POP constraints as follows:

$$\mathbf{E}_i^{\text{bpop+}} = \left\{ E_{i,t}^{\text{bpop+}} \mid E_{i,t}^{\text{bpop+}} = E_i^{\prime\text{left}}, \forall t \in \left\{0, \ldots, T_i^{\text{left}}\right\} \right\}, \tag{6.22}$$

$$\mathbf{E}_i^{\text{bpop-}} = \left\{ E_{i,t}^{\text{bpop-}} \mid E_{i,t}^{\text{bpop-}} = \max(E_i'^{\text{left}} - P_i'^{\max}(T_i^{\text{left}} - t),\ 0), \right.$$
$$\left. \forall t \in \left\{0, \ldots, T_i^{\text{left}}\right\} \right\}, \tag{6.23}$$

$$\forall i \in \mathscr{F}_{c, s_{\text{cur}}},$$

where T_i^{left} is the total number of control periods left (e.g., $T_i^{\text{left}} = 3$ in Fig. 6.7).

Once an EV has constructed its binned POP constraints, these constraints are classified in the right bin. Each bin is characterized based on two parameters: T_i^{left} and T_i^{slope}. Based on these parameters, the set of connected EVs $\mathscr{F}_{c,s_{\text{cur}}}$ has a partition $\mathscr{P}_{\text{bin}} = (B_1, \ldots, B_{|\mathscr{P}_{\text{bin}}|})$ defined as:

$$\forall n \in \{1, \ldots, |\mathscr{P}_{\text{bin}}|\},$$
$$T_i^{\text{left}} = T_j^{\text{left}}, \qquad \forall i, j \in B_n, \tag{6.24}$$
$$T_i^{\text{slope}} = T_j^{\text{slope}}, \qquad \forall i, j \in B_n. \tag{6.25}$$

In a single bin, the binned POP constraints and maximum power of its contained EVs are then summed:

$$\forall n \in \{1, \ldots, |\mathscr{P}_{\text{bin}}|\},$$
$$\mathbf{E}_n^{\text{bin+}} = \sum_{i \in B_n} \mathbf{E}_i^{\text{bpop+}}, \tag{6.26}$$
$$\mathbf{E}_n^{\text{bin-}} = \sum_{i \in B_n} \mathbf{E}_i^{\text{bpop-}}, \tag{6.27}$$
$$P_n^{\text{bin}} = \sum_{i \in B_n} P_i^{\max}. \tag{6.28}$$

A solution within these constraints is then calculated in the optimization mechanism (Sect. 6.4.3):

$$\mathbf{E}_n^{\text{chrg}} = \{E_{n,1}^{\text{chrg}}, \ldots, E_{n,T_n^{\text{bin}}}^{\text{chrg}}\}. \tag{6.29}$$

The path is then translated into a relative schedule for each EV in a bin as follows:

$$\boldsymbol{\alpha}_n^{\text{bin}} = \{\alpha_{n,t}^{\text{bin}} \mid \alpha_{n,t}^{\text{bin}} = \frac{E_t^{\text{path}} - E_{n,t}^{\text{bin-}}}{E_{n,t}^{\text{bin+}} - E_{n,t}^{\text{bin-}}},\ \forall t \in \{1, \ldots, T_n^{\text{bin}}\}\}. \tag{6.30}$$

Individual EVs will use this relative schedule in their own regulation bounds to determine individual GIV schedules:

$$\mathbf{E}_i^{\text{path}} = \{E_{i,t}^{\text{path}} \mid E_{i,t}^{\text{path}} = E_{n,t}^{\text{bpop-}} + \alpha_{i,t}^{\text{bin}}(E_{n,t}^{\text{bpop+}} - E_{n,t}^{\text{bpop-}}), \quad \forall t \in \{1, \ldots, T_n^{\text{left}}\}\}. \tag{6.31}$$

This mechanism is much faster than calculating individual paths for each EV and it is assured that the sum of the individual POP schedules amounts to the total POP schedule.

Using this binning mechanism, the EV can continuously update its local schedule. When a new control period begins, the first step in the schedule is used as the requested POP in the MDP in (6.11).

6.4.3 Scheduling Mechanism

The goal of this step is to determine a POP schedule for each bin that maximizes the number of regulation bids in the market. To achieve this goal, the aggregator must submit new bids while providing enough regulation capacity for bids that are already submitted. The problem for finding a POP schedule is formulated as a lexicographic multi-objective optimization problem [12]. In this type of multi-objective optimization, objectives are ordered based on their priority, with a set of solutions first found for the objective with the highest priority. This set of solutions is then further reduced by optimizing for the second objective and so on. To clarify each step in our multi-objective optimization, the descriptions in this section are interlaced with the example shown in Fig. 6.8.

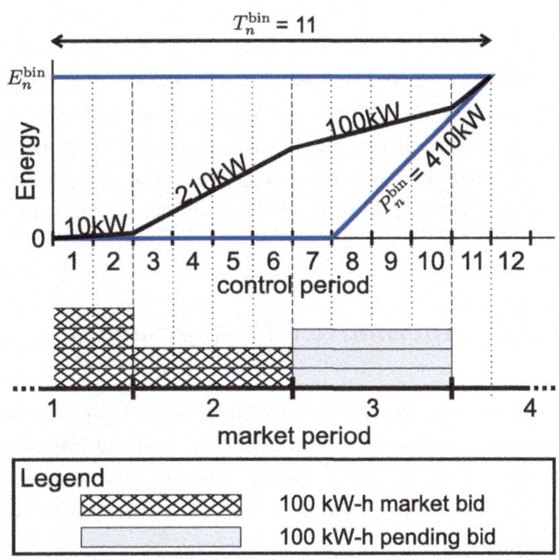

Fig. 6.8 Visualization of the aggregator's GIV scheduling problem

Figure 6.8 shows a scheduling example with one bin B_1. In this bin, 410 kWh must be scheduled at a maximum power of 410 kW in 11 control steps. The upper graph shows these parameters as POP schedule constraints and the lower graph shows hourly regulation bids with their contained regulation capacity. In this example, the regulation market accepts bids with a granularity of 100 kWh.

The first objective is to provide regulation capacity for bids already submitted to the regulation market. This objective has the highest priority because submitted bids cannot be withdrawn. In our example, a 400-kWh bid was submitted in the first market timeslot and a 200-kWh bid was submitted in the second market timeslot. To comply with these bids, POP values of 10 and 210 kW are scheduled in these first timeslots. The second objective is to maximize the number of future bids that can be submitted to the market. In the example, the only market timeslot eligible for future bids is timeslot 3 because the binned POP constraints end before timeslot 4. The third objective is to provide additional regulation capacity for pending bids to impose a safety margin on the unpredictable regulation signal received from the aggregator, which influences regulation capacity. The fourth objective is to flatten the EV fleet's total load to avoid the high costs and infrastructure degradation associated with peak loads.

The aggregator continuously performs this multi-objective optimization based on the most recent information in different bins. This is necessary because EVs are continuously being plugged in and unplugged, the regulation signal influences individual SOCs, and so on. Just before a market period begins, the aggregator will submit a bid based on the latest POP schedule. Each time a new POP schedule is calculated, this schedule is dispatched between EVs as shown in (6.30).

6.5 Evaluation

In this section, the proposed GIV approach is evaluated through benchmarking against centralized and decentralized GIV approaches. A centralized approach guarantees an optimal GIV schedule, whereas a distributed approach guarantees optimal scalability in the absence of a central computation unit. The goal of this evaluation is to demonstrate that the proposed GIV approach achieves a near-optimal GIV schedule while providing sufficient scalability for large EV fleets.

6.5.1 Alternative GIV Approaches

This section presents both alternative GIV approaches used as benchmarks in our evaluation. These benchmarks are designed by adjusting the models and interactions proposed in Sect. 6.4. In the rest of this section, the differences between the bin-based GIV approach and the two alternative GIV approaches is discussed and formalized.

A centralized GIV approach guarantees optimal POP schedules for each EV. Consequently, instead of approximating the POP constraints as in (6.22) and (6.23), this approach uses the original POP constraints shown in (6.19) and (6.20). To further guarantee optimality, these POP constraints are not aggregated. Instead, each POP constraint is classified in its own bin, i.e., $N_{\text{bin}} = N_{\text{ev}}$, and the set of EVs is partitioned as $\mathscr{P}_{\text{bin}} = (\{1\}, \{2\} \ldots, \{N_{\text{ev}}\})$. This allows the aggregator to always calculate an individual POP schedule for each EV in its fleet.

On the other hand, a distributed benchmark guarantees optimal scalability. Instead of centrally calculating a POP schedule for each EV, each EV determines its own POP schedule by myopically maximizing regulation capacity [3, 13, 14]. In the simplest form of this GIV schedule, each EV sends only its expected regulation capacity to the aggregator. To achieve the same behavior in the models and interactions proposed in this work, the original POP constraints in (6.19) and (6.20) are used, and each EV is classified in its own bin. However, unlike the centralized GIV approach, no optimization is performed and the aggregator sends a relative schedule in (6.32) to each bin with only zeros instead:

$$\boldsymbol{\alpha}_n^{\text{bin}} = \{\alpha_{n,t}^{\text{bin}} \mid \alpha_{n,t}^{\text{bin}} = 0, \ \forall t \in \{1, \ldots, T_n^{\text{bin}}\}\}. \tag{6.32}$$

The result of using this schedule is that the EVs charge at a POP of 0 kW for as long as possible, until the slope of the regulation boundary is reached and the EV needs to be charged at maximum power. Consequently, at a POP of 0 kW, the EV myopically provides the maximum regulation capacity until the battery must be charged and regulation capacity is reduced to 0 kWh.

In summary, the centralized GIV approach calculates individual POP schedules for each EV based on their original POP constraints, and the distributed GIV approach myopically maximizes regulation capacity. Both alternatives can be integrated in the proposed model, which allows an in-field system to easily switch between the three solutions. In the rest of the section, the three solutions are compared in terms of optimal scheduling and scalability.

6.5.2 *Simulation Results: Scheduling Quality*

To evaluate the GIV approach in terms of optimization objectives, all three GIV control approaches are evaluated in simulations based on the operational EV fleet at the University of Delaware [15]. In this EV fleet, staff members use EVs to commute between their homes and the university. Each EV has a maximum power transfer of 12 kW and a battery capacity of 35 kWh. A scaled-up version of 120 EVs is simulated based on these parameters. Each EV can be charged at home and the university, and departure, arrival, and driving times are drawn from a set of artificial probability distributions. Departure times from home in the morning are sampled from a normal distribution $N(\mu_{\text{dept}} = 8\text{ am}, \sigma_{\text{dept}} = 2\text{ h})$ and morning driving

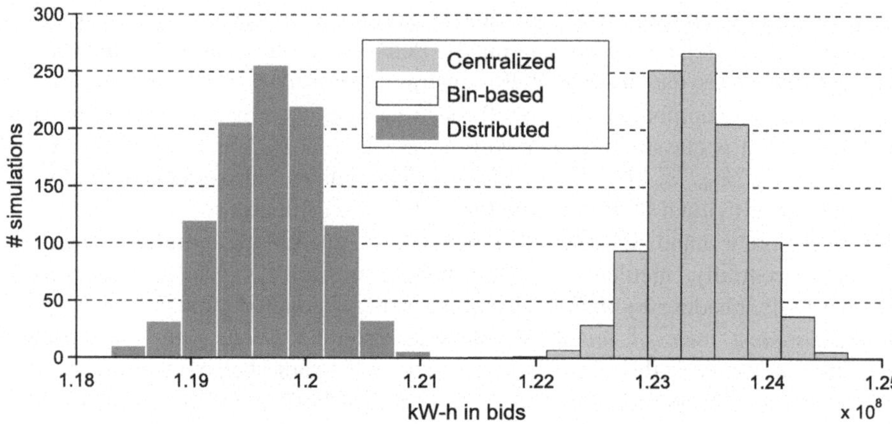

Fig. 6.9 Distribution of the total regulation capacity in all bids for each solution as result of a Monte-Carlo experiment

times are sampled from $N(\mu_{\text{drive}} = 30 \text{ min}, \sigma_{\text{drive}} = 15 \text{ min})$. Similarly, departure times from work are sampled from $N(\mu_{\text{dept}} = 5 \text{ pm}, \sigma_{\text{dept}} = 2 \text{ h})$, and driving times are sampled from $N(\mu_{\text{drive}} = 30 \text{ min}, \sigma_{\text{drive}} = 15 \text{ min})$.

The GIV approach in the first simulation is benchmarked in terms of optimization objectives. The primary objective is maximizing bids, and the secondary objective is flattening the load. These objectives are analyzed in a simulation of 120 EVs over a period of 5 days (one preceding day with artificial SOCs is simulated and discarded). Each solution is simulated 1000 times with different driving cycles sampled from the same distributions. Figure 6.9 shows the distributions of the total amount of bids resulting from each solution plotted in a histogram. The left end of the histogram represents the distributed GIV approach and right end represents the bin-based GIV approach and centralized GIV approach. The difference between the distributed and bin-based GIV approaches is $\sim 3\%$. Although all three GIV approaches are relatively close to each other in terms of bids, the decentralized GIV approach is clearly outperformed in terms of load variation (Fig. 6.10). Additionally, although both the bin-based and centralized GIV approaches achieve a similar load profile, the distributed approach nearly doubles the peak load.

6.5.3 Simulation Results: Scheduling Scalability

To evaluate this GIV approach in terms of computational scalability, all three GIV control approaches are evaluated in a variety of situations based on the base scenario presented in previous section.

The first experiment analyzes the execution time for solving a scheduling problem at the aggregator (Fig. 6.11). The input to the aggregator's scheduling problem, which indicates the scale of the problem, is the number of EVs. Increasing the

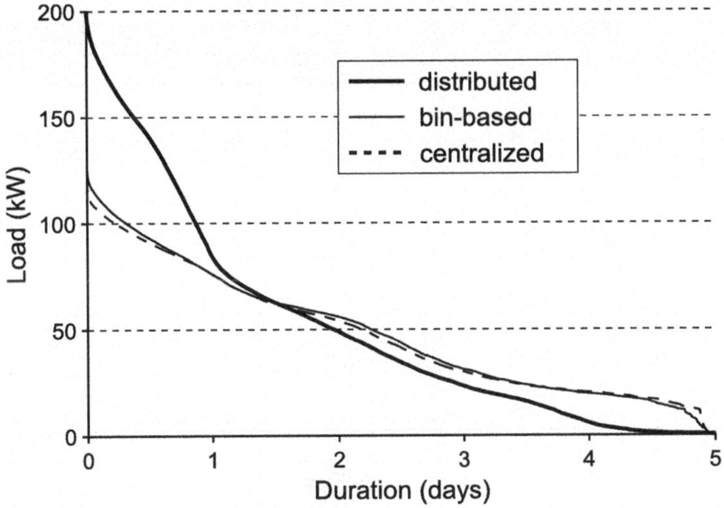

Fig. 6.10 Load duration curve of all three GIV approaches

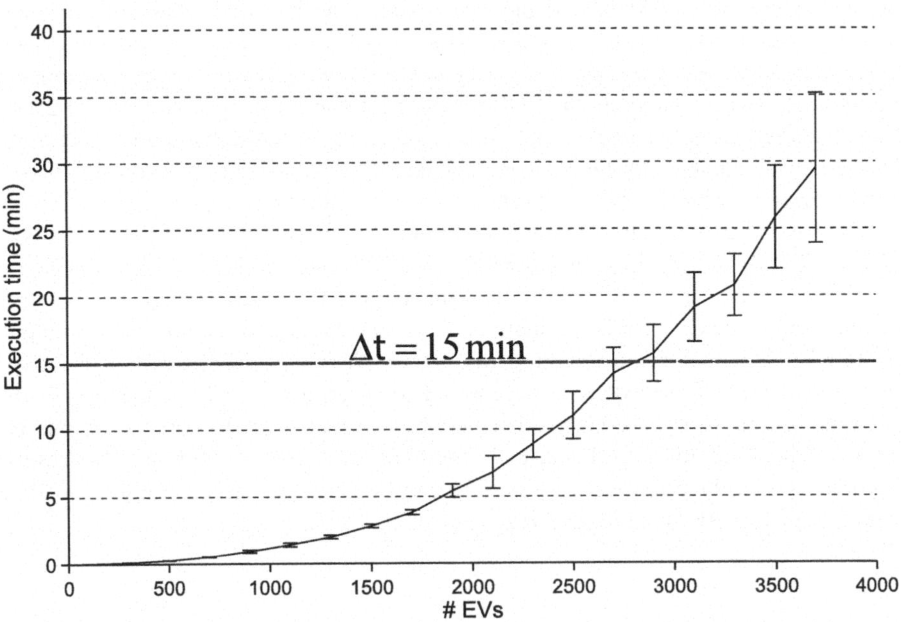

Fig. 6.11 Computational scalability analysis

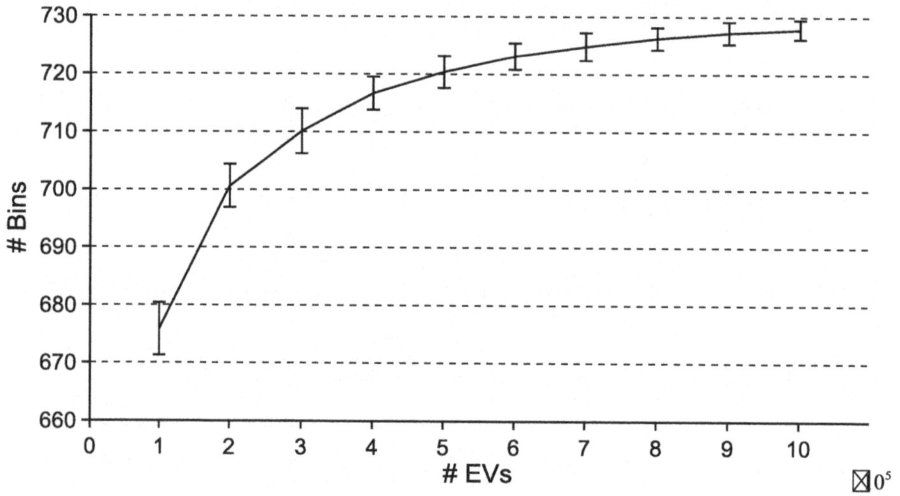

Fig. 6.12 Number of bins resulting from a varying number of EVs

number of EVs from 100 to 3700 increases execution time from a few milliseconds to half an hour. A critical boundary is the length of a control period, which is reached at approximately 2800 EVs. Beyond this border, the aggregator no longer has enough time to calculate the POP for the next control period. At close to 2800 EVs, the aggregator has just enough time to calculate a POP schedule for each EV. However, any changes in the environment such has changing regulation signals or newly arrived EVs will not be considered during calculation of a new schedule.

The second experiment calculates the number of bins for an increasing number of EVs (Fig. 6.12). A bin is characterized by T^{left}, the number of control periods left before departure, and T^{slope}, the time required to fully charge the battery at maximum power. In this evaluation, EVs can charge to a maximum amount of 35 kWh at a maximum charging rate of 12 kW. Consequently, a conservative maximum of 12 control periods is required. Furthermore, for EVs charging both at home and at work, a conservative maximum of 24 h or 96 control periods is chosen. This amounts to a theoretical maximum of a mere 1416 bins. This result is confirmed by the empirical results shown in Fig. 6.12, where the number of EVs varies from 100,000 to 1,000,000 EVs.

6.6 Conclusions and Future Work

In this paper, we studied an aggregator's problem of providing regulation services with many EVs. To solve this problem, a bin-based GIV control approach is proposed. In this approach, EVs with similar charging characteristics are classified in bins, and rather than calculating a charging schedule for each EV, a schedule is calculated for each bin, which significantly reduces the scheduling problem's computational complexity. In a simulation, this GIV approach was found to achieve a similar solution quality to a centralized GIV approach while achieving good scalability qualities. Even for millions of EVs, the bin-based GIV approach can solve the optimization problem in a limited amount of time.

Future and ongoing work focuses on using bin scheduling for different scenarios such as time of use (TOU) tariffs, day-ahead markets, and peak power provision [16]. Furthermore, the aggregator's optimization problem will be further extended to include predictions and stochastic data.

References

1. R.J. Bessa, M.A. Matos, The role of an aggregator agent for EV in the electricity market, in *7th Mediterranean Conference and Exhibition on Power Generation, Transmission, Distribution and Energy Conversion (MedPower 2010)* (2010), pp. 1–9. https://doi.org/10.1049/cp.2010.0866
2. D. Wu, D. Aliprantis, L. Ying, Load scheduling and dispatch for aggregators of plug-in electric vehicles. IEEE Trans. Smart Grid **3**(1), 368–376 (2012)
3. S. Kamboj, W. Kempton, K.S. Decker, Deploying power grid-integrated electric vehicles as a multi-agent system, in *The 10th International Conference on Autonomous Agents and Multiagent Systems - Volume 1, AAMAS'11* (2011), pp. 13–20
4. E. Sortomme, M.A. El-Sharkawi, Optimal scheduling of vehicle-to-grid energy and ancillary services. IEEE Trans. Smart Grid **3**(1), 351–359 (2012)
5. S.-L. Andersson, A.K. Elofsson, M.D. Galus, L. Göransson, S. Karlsson, F. Johnsson, G. Andersson, Plug-in hybrid electric vehicles as regulating power providers: case studies of Sweden and Germany. Energy Policy **38**(6), 2751–2762 (2010)
6. E. Sortomme, M.A. El-Sharkawi, Optimal charging strategies for unidirectional vehicle-to-grid. IEEE Trans. Smart Grid **2**(1), 131–138 (2011)
7. A.T. Al-Awami, E. Sortomme, Coordinating vehicle-to-grid services with energy trading. IEEE Trans. Smart Grid **3**(1), 453–462 (2012)
8. S. Koch, J.L. Mathieu, D.S. Callaway, Modeling and control of aggregated heterogeneous thermostatically controlled loads for ancillary services, in *17th Power System Computation Conference* (2011)
9. P. Codani, M. Petit, Y. Perez, Diversity of transmission system operators for grid integrated vehicles, in *2014 11th International Conference on European Energy Market (EEM)* (2014), pp. 1–5
10. Dispatch PJM (PJM), Manual 12: Balancing Operations Revision: 39 (2019). https://www.pjm.com/~/media/documents/manuals/m12.ashx
11. J. Tomić, W. Kempton, Using fleets of electric-drive vehicles for grid support. J. Power Sources **168**(2), 459–468 (2007)
12. R.T. Marler, J.S. Arora, Survey of multi-objective optimization methods for engineering. Struct. Multidiscip. Optim. **26**(6), 369–395 (2004)
13. S. Vandael, T. Holvoet, G. Deconinck, S. Kamboj, W. Kempton, A comparison of two GIV mechanisms for providing ancillary services at the University of Delaware, in *2013 IEEE International Conference on Smart Grid Communications (SmartGridComm)* (2013), pp. 211–216

14. S. Kamboj, K. Decker, K. Trnka, N. Pearre, C. Kern, W. Kempton, Exploring the formation of electric vehicle coalitions for vehicle-to-grid power regulation, in *Proceedings of 1st International Workshop on Agent Technologies for Energy Systems (ATES-2010)* (2010), pp. 67–74
15. University of Delaware, The grid-integrated vehicle with vehicle to grid technology (2014). http://www.udel.edu/V2G/
16. W. Kempton, T. Kubo, Electric-drive vehicles for peak power in Japan. Energy Policy **28**(1), 9–18 (2000)

Chapter 7
A Continuum Approach to Assessing the Impact of Spatiotemporal EV Charging to Distribution Grids

Yoshihiko Susuki, Naoto Mizuta, Akihiko Kawashima, Yutaka Ota, Atsushi Ishigame, Shinkichi Inagaki, and Tatsuya Suzuki

7.1 Introduction

The design of coordinated use of electric vehicles (EVs) is important for future integrated management of transportation and energy. Examples of integration include an *EV-sharing system* [1, 2] that can track trajectories and battery status to satisfy users requests. Kawashima et al. [3] proposed using an EV-sharing system for community-level management of electrical energy. This work opens a new possibility for integrated transportation-energy management, in which the EV-sharing system partly contributes to the power grid. Especially, management of charging/discharging of a large population of EVs is expected to provide a control means for the power grid. The idea is generally known as *ancillary service* (AS) [4] provided by EVs and has attracted significant interest in recent years, e.g., [5–10].

Y. Susuki (✉)
Graduate School of Engineering, Osaka Prefecture University, Sakai, Osaka, Japan
e-mail: susuki@eis.osakafu-u.ac.jp

N. Mizuta · A. Ishigame
Department of Electrical and Information Systems, Osaka Prefecture University, Sakai, Osaka, Japan

A. Kawashima
Department of Mechanical Science and Engineering, Nagoya University, Nagoya, Japan

Y. Ota
Power System Design Laboratory, Tokyo City University, Tokyo, Japan

S. Inagaki · T. Suzuki
Department of Mechanical Systems Engineering, Nagoya University, Nagoya, Aichi, Japan

T. Suzuki et al. (eds.), *Design and Analysis of Distributed Energy Management Systems*, Power Electronics and Power Systems,
https://doi.org/10.1007/978-3-030-33672-1_7

The coordinated use of EVs poses several problems in terms of dynamic performance of the power grid. One problem is to assess how spatiotemporal charging (and possibly discharging) of EVs impacts on voltage profile of a power distribution grid. An EV is considered as an autonomously moving battery in the spatial domain and can, in principle, conduct charging/discharging *anywhere* in the grid. Thus, assessing the impact of EV charging on distribution voltage is challenging. In fact, Clement-Nyns et al. [11, 12] studied the impact of EV charging on the distribution grid using load-flow analysis, i.e., a conventional space-discretization approach based on power-flow equations. Additionally, in the OECD/IEA's Nordic EV Outlook [13], it was suggested that "in Norway, a few percent of the transformers are already overloaded and estimates that an increase in average power consumption of 1–2 kW household would lead to overload in 10% of the transformers." This increase is possibly due to the charging of EVs. In the case of EV-sharing which we now address, it is often planned that a large population of EVs would be distributed in a relatively small area, referred to as the *last mile transportation* within a range of a few kilometers [2]. Thus many charging stations can be densely placed in a small area, possibly causing a non-negligible impact on distribution voltage.

The purpose of this chapter is to demonstrate an alternative approach for solving the assessment problem by using a recently derived continuum representation of distribution voltage profile—nonlinear ordinary differential equation (ODE) [14, 15].The ODE representation explicitly keeps spatial information on a distribution grid and thus enables us to quantify spatial impact on distribution voltage, such as how far EV charging at a particular location affects the voltage profile. Baek et al. [16] and Susuki et al. [17] reported preliminary studies on this approach, in particular, numerical implementation and algorithms for solving the nonlinear ODE. In this chapter, following those studies [16, 17], we demonstrate this approach through numerical simulations of a simple distribution model incorporated with operational data on EVs in a EV-sharing system demonstration project in Japan. The main part of this chapter is based on the work of Mizuta et al. [18]. Such assessment of dynamic performance in coupled transportation and electrical systems through continuum representation is direct to the next design research on integrated infrastructure [19, 20]. In fact, in companion papers [21, 22], we proposed using the nonlinear ODE to design a coordination of EVs to provide AS. The contents of this chapter are reported in our conference proceeding [23].

The rest of this chapter is organized as follows. In Sect. 7.2, we introduce ODE modeling of a distribution voltage profile. In Sect. 7.3, we demonstrate how the nonlinear ODE is integrated with operational data on EVs for assessing the impact of EV charging. Section 7.4 is devoted to discussions related to the demonstration, and Sect. 7.5 is the conclusion of this chapter with a brief summary.

7.2 ODE Modeling of Distribution Voltage Profile

First, we introduce ODE modeling of distribution voltage profile based on the work of Chertkov et al. [14]. To simplify the introduction, we assume that no voltage regulation devices such as transformers or step voltage regulator (SVR) are operated. Thus, we can consider the voltage profile starting at a distribution substation (bank) that is continuous in space (length). We will mention how to include the effects of voltage regulation devices in the ODE representation later, but for now, We consider a single distribution feeder as shown in Fig. 7.1, starting at a bank where we introduce the origin of the one-dimensional displacement (location) $x \in \mathbb{R}$ as $x = 0$. The voltage phasor at location x is represented with $v(x)\exp\{\mathrm{i}\theta(x)\}$, where i denotes an imaginary number, $v(x)$ the *voltage amplitude* [V], and $\theta(x)$ the *voltage phase* [rad]. Also, as new functions in x, we define the *power transfer density* [V^2/m] at x by $s(x) := -v(x)^2 d\theta(x)/dx$ and the *voltage gradient* [V/m] by $w(x) := dv(x)/dx$. Thus, the four functions θ, v, s, and w are related to the following first-order nonlinear ODE [14]:

$$\left.\begin{aligned}
\frac{d\theta}{dx} &= -\frac{s}{v^2} \quad (v \neq 0),\\
\frac{dv}{dx} &= w,\\
\frac{ds}{dx} &= \frac{b(x)p(x) - g(x)q(x)}{g(x)^2 + b(x)^2},\\
\frac{dw}{dx} &= \frac{s^2}{v^3} - \frac{g(x)p(x) + b(x)q(x)}{v\{g(x)^2 + b(x)^2\}}.
\end{aligned}\right\} \tag{7.1}$$

The functions $g(x)$ and $b(x)$ in (7.1) are the position-dependent conductance and susceptance per-unit-length, and both are assumed to be continuous in x. Additionally, the functions $p(x)$ (or $q(x)$) are the active (or reactive) power flowing from the feeder (note that $p(x) > 0$ indicates positive active-power flowing to the feeder at x). We will call $p(x)$ and $q(x)$ the *power density functions* and model them using Dirac's delta function for analytical treatment in Sects. 7.3.2 and 7.4.2, and with a Gaussian function with a small variance for numerical demonstration in Sect. 7.3.2.

To solve the nonlinear ODE (7.1), it is necessary to determine the functions $g(x)$ and $b(x)$, the power density functions $p(x)$ and $q(x)$ as well as a *boundary*

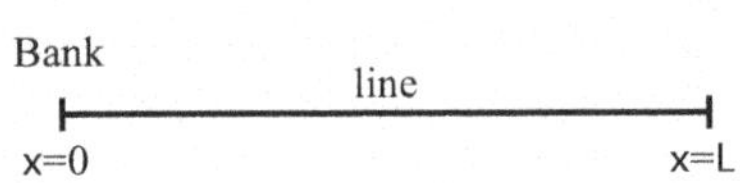

Fig. 7.1 A single distribution feeder that starts at a bank, goes through a finite-length line, and ends at a non-loading terminal

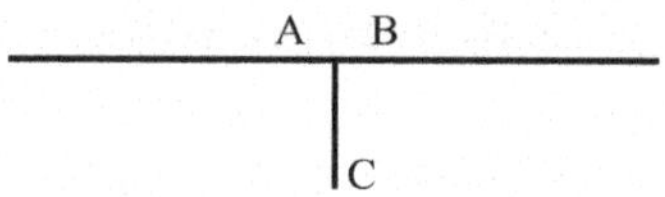

Fig. 7.2 A simple model of a distribution feeder with a bifurcation point that connects three different lines labeled as A, B, and C

condition. Consider again the single distribution feeder in Fig. 7.1. Because of voltage regulation at the bank at the starting point $x = 0$, we naturally set $v(0)$ and $\theta(0)$ as constants. In this chapter, $v(0)$ coincides with unity in a per-unit system and $\theta(0)$ coincides with zero as a reference. At the end point $x = L$ (i.e., no feeder or load exist at $x > L$), we derive the conditions $s(L) = 0$ and $w(0) = 0$ supposing the end is not loaded [16, 17]. The boundary conditions at the starting and end points can thus be summarized as:

$$\theta(0) = 0, \quad v(0) = 1, \quad s(L) = 0, \quad w(L) = 0.$$

Lastly, to consider a distribution feeder with a bifurcation, it is also necessary to determine the boundary condition at the bifurcation point. A simple model with one bifurcation point connecting three lines labeled A, B, and C is shown in Fig. 7.2. By direct calculation using (7.1), the following boundary condition at the bifurcation point is derived in [17] as:

$$\theta_{\mathrm{A}} = \theta_{\mathrm{B}} = \theta_{\mathrm{C}}, \quad v_{\mathrm{A}} = v_{\mathrm{B}} = v_{\mathrm{C}}, \quad s_{\mathrm{A}} = s_{\mathrm{B}} + s_{\mathrm{C}}, \quad w_{\mathrm{A}} = w_{\mathrm{B}} + w_{\mathrm{C}},$$

where θ_{A}, v_{B}, and so on represent dependent variables taken as bifurcation point limits to along the lines A, B, and C, respectively.

The boundary condition allows incorporation of the effects of voltage regulation devices such as transformers and SVRs. Their operations are regarded as *discontinuous* changes of voltage amplitude where they are installed. Suppose that in Fig. 7.1, a tap-changing transformer with voltage ratio n is installed at the location $x = \xi$. Then, assuming zero power density at $x = \xi$ (namely, $p(\xi) = q(\xi) = 0$), we obtain

$$\left.\begin{aligned} \theta(\xi-) = \theta(\xi+),\ n \cdot v(\xi-) = v(\xi+), \\ s(\xi-) = s(\xi+),\ \ w(\xi-) = n \cdot w(\xi+) \end{aligned}\right\}$$

where $\theta(\xi-)$, $v(\xi+)$, and so on represent dependent variables taken as limits to the transformer location along the line, i.e., $\theta(\xi-) = \lim_{\delta \nearrow 0} \theta(\xi + \delta)$ and $v(\xi+) = \lim_{\delta \searrow 0} v(\xi + \delta)$.

7.3 Numerical Demonstration

Next, we discuss how to exploit the nonlinear ODE (7.1) to assess the impact of EV charging on the distribution voltage profile. In this section, we use a simple feeder model and numerical simulations to demonstrate how the idea works by integrating the nonlinear ODE with operational data on EVs.

7.3.1 Setting of Feeder and Charging Stations

Here, we investigate distribution feeder and charging stations model shown in Fig. 7.3. The feeder includes one bifurcation point and has 16 charging stations spaced at a 1-km interval. The bifurcation point exists in the middle between the stations labeled 4 and 5. We assume that the two ends of the feeder have no charging stations to simplify the current analysis. Additionally, to focus on the impact of EV charging, it is assumed that the system has no generation facility or load work aside from the EVs. The constant voltage at bank is set at 6.6 kV, and its rated capacity at 40 MVA. Furthermore, it is assumed that each feeder has common and constant conductance and susceptance in x:

$$g(x) = G, \quad b(x) = B.$$

The feeder's resistance (or reactance) is set to 0.227 Ω/km (or 0.401 Ω/km) and the ratio of resistance to reactance (identically, G/B) is 0.566. These values are based on the standard settings of medium-voltage distribution grids in Japan. In the following, we use a per-unit system [24] for numerical simulations of the distribution feeder model.

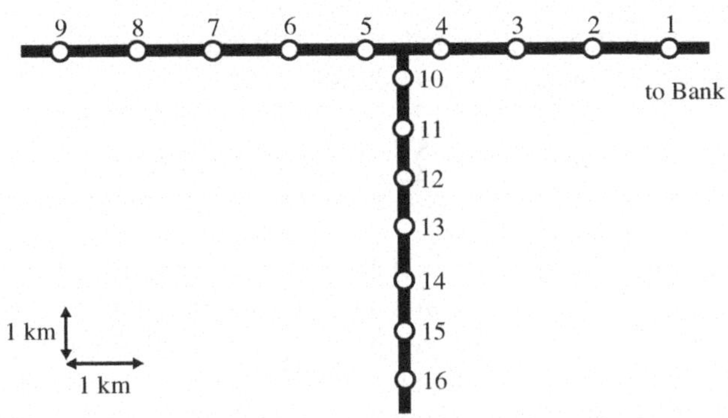

Fig. 7.3 Settings of distribution feeder and charging stations for the numerical demonstration. The uniform feeder includes one bifurcation point and has 16 charging stations spread at a 1-km interval. The two terminals are assumed to have no load for simplicity of the current analysis

7.3.2 Construction of Power Density Function

Next, we consider power density functions $p(x)$ and $q(x)$. Basically, active-power $p(x)$ is decomposed into the three components as follows (the same holds for reactive-power $q(x)$):

$$p(x) = p_{\mathrm{G}}(x) + p_{\mathrm{L}}(x) + p_{\mathrm{EVs}}(x). \tag{7.2}$$

The first term in the right-hand side of (7.2), $p_{\mathrm{G}}(x)$, represents the contribution of generation facilities such as PV panels and cogeneration units, which are not considered in the following simulations. Power conditioning systems (PCS), such as grid-connected inverters for distributed power sources, and advanced technology, such as smart inverters [25], are considered in this term and the associated reactive-power term $q_{\mathrm{G}}(x)$. Additionally, it is possible to consider time-varying generation as $p_{\mathrm{G}}(x,t)$. In this role, (7.1) becomes a partial differential equation that governs spatiotemporal changes of quantities such as $v(x,t)$.

The second term in (7.2), $p_{\mathrm{L}}(x)$, represents the contribution of loads. Although this term is not used in the following simulations, it benefits general use explaining how load contributions can be modeled. The polynomial or ZIP model is the sum of the three major static load types [24]: constant impedance, constant current, and constant power. Here, we introduce three position-dependent functions of loads that are constant in *time*: resistance $r_{\mathrm{C.L}}(x)(\neq 0)$, current $i_{\mathrm{C.L}}(x)$, and active-power $p_{\mathrm{C.L}}(x)$. Thus, it is natural to incorporate the ZIP model into the term $p_{\mathrm{L}}(x)$ as:

$$p_{\mathrm{L}}(x) = \frac{\{v(x)\}^2}{r_{\mathrm{C.L}}(x)} + v(x)i_{\mathrm{C.L}}(x) + p_{\mathrm{C.L}}(x).$$

The same incorporation holds for reactive-power $q_{\mathrm{C.L}}(x)$. Note that time-varying loads can be incorporated like $i_{\mathrm{C.L}}(x,t)$. A dynamic model with $v(x,t)$ as input and $i_{\mathrm{C.L}}(x,t)$ as output can be also incorporated.

The last term $p_{\mathrm{EVs}}(x)$ represents the contribution of EV charging operations through a station. By denoting the index set of charging station locations as $\mathscr{I}_{\mathrm{Sta}}$, we can formulate the term $p_{\mathrm{EVs}}(x)$ as follows:

$$p_{\mathrm{EVs}}(x) = \sum_{i \in \mathscr{I}_{\mathrm{Sta}}} P_{\mathrm{EVs}}[i]\delta(x - \xi_i),$$

where $\delta(x - \xi_i)$ denotes the Dirac's delta function supported at $x = \xi_i$, and ξ_i represents the spatial location (coordinate) of charging station i. Here, $P_{\mathrm{EVs}}[i]$ represents the amount of charging power at station i and can change over *time*. More precisely, it depends on the number of EVs that arrive at the station and stay there under charging operation. For simplicity of the notation, we do not express the time dependency of $P_{\mathrm{EVs}}[i]$ explicitly.

In this chapter, we construct an approximation of the term $p_{\mathrm{EVs}}(x)$ based on practical data and a coarse-graining technique. Our group was provided with

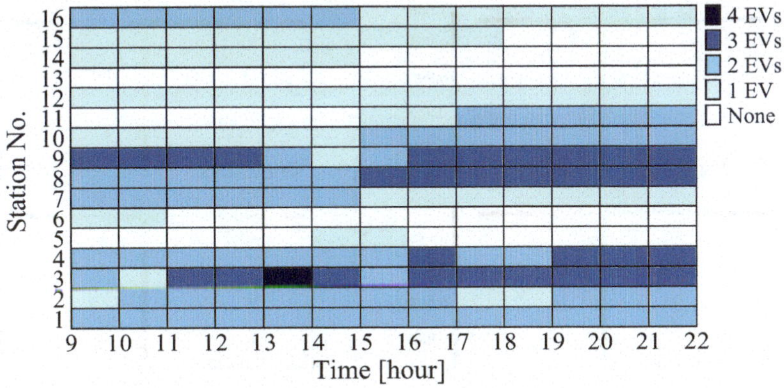

Fig. 7.4 Spatiotemporal pattern of several EVs at the 16 charging stations in the distribution feeder shown in Fig. 7.3. Multiple EVs arrive at, stay at, or leave the charging stations. The pattern was fixed with operational data on EVs from an EV-sharing system demonstration project in Japan

operational data on EVs in an EV-sharing demonstration project. With this practical data, we fixed a spatial temporal pattern of several EVs arriving at the 16 charging stations, which is shown in Fig. 7.4. Multiple EVs arrive, stay, or leave the 16 charging stations in this pattern. Additionally, to avoid the difficulty of numerical simulations due to Dirac's delta function in $p_{\mathrm{EVs}}(x)$, we used the following coarse-graining method for the term using a Gaussian function [16]:

$$p_{\mathrm{EVs}}(x) \sim \sum_{i \in \mathscr{I}_{\mathrm{Sta}}} \frac{P_{\mathrm{EVs}}[i]}{\sqrt{2\pi\sigma^2}} \exp\left(-\frac{(x-\xi_i)^2}{2\sigma^2}\right),$$

where we regard x and ξ_i as scalars and σ^2 is the variance. For numerical simulations we assumed that any EV staying at a station was under charging operation, and we fixed a single EV charging power at 4 kW based on Clement-Nyns et al. [11]. Also, to vividly exhibit the impact of EVs, we used the function $p_{\mathrm{EVs}}(x)$ multiplied by 10 in the simulations. We fixed variance σ at 0.05 km, which is much smaller than the interval between stations (1 km). Reactive-power $q(x)$ was negligible here because EV charging operations under a unity power-factor are normal in current practice.

7.3.3 *Results*

Figure 7.5 shows snapshots of distribution voltage amplitude and the gradient of the distribution feeder model at 9:00 and 18:00. The results are solutions of the nonlinear ODE (7.1) numerically obtained using Euler's discretization and an iterative scheme: see Susuki et al. [17] for details on the scheme. It is seen from

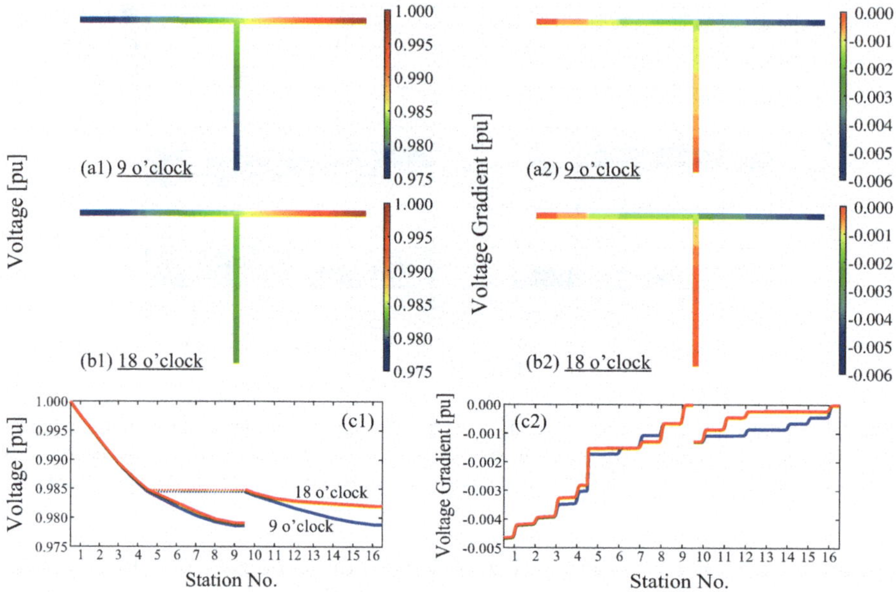

Fig. 7.5 Snapshots of distribution voltage profile and voltage gradient driven by spatiotemporal EV charging at 9:00 and 18:00. The voltage profile changes with EV charging operations as shown in Fig. 7.4. The voltage gradient vividly shows dependence on the number of EVs. Note that the horizontal axis of the figures (**c1**, **c2**) denotes the location of charging stations involving information on displacement x. ©IEEE

Fig. 7.4 that at 9:00, the number of EVs staying at each station is relatively flat. Thus, in Fig. 7.5a1, c1, the voltage amplitude gradually decreases to the ends of the feeder. Additionally, in Fig. 7.5a2, c2, we see that the voltage gradient exhibits large changes at the stations. Next, at 18:00 in Fig. 7.4, the number of EVs is not flat for every station. In particular, the number of EVs at stations 13–16 is lower than at 9:00. Thus, for the part of the feeder with stations 10–16, the decrease of voltage amplitude at 18:00 is smaller than at 9:00. This is clearly visualized in Fig. 7.5a2, b2, c2 where the voltage gradients at 9:00 and 18:00 behave differently.

Here we look in detail at the relationship between power density function change and voltage gradient one. An example of the relationship is presented in Fig. 7.6. In Fig. 7.4, from 9:00 to 10:00, the number of EVs at station 2 (or 3) increases (or decreases) by one. Based on this, in Fig. 7.6a, the power density function is locally re-shaped at stations 2 and 3. Thus, in Fig. 7.6b, a slight but global change in voltage amplitude is observed on the feeder line after station 2. For the same situation, as shown in Fig. 7.6c, the voltage gradient on the feeder changes between stations 2 and 3 locally. This observation implies that EV charging causes both local voltage gradient changes and global voltage amplitude changes. The voltage gradient is a unique physical quantity derived from the continuum representation and is regarded as a hidden variable controlling distribution voltage.

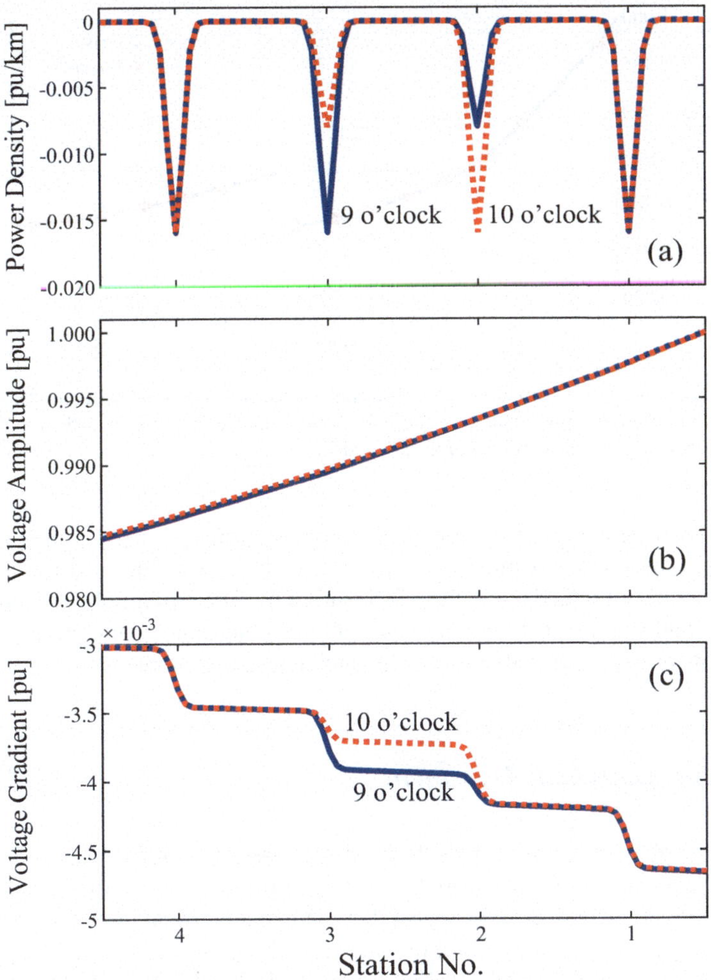

Fig. 7.6 Example of the relationship of changes in the power density function, voltage amplitude, and voltage gradient. Changes in the number of EVs at station 2 and 3 cause the local change of voltage gradient and the slight global change of voltage amplitude. ©IEEE

7.4 Discussions

7.4.1 Comparison with Power-Flow Equation

Distribution voltage profiles have been studied using the power-flow equation [24]. Here we report a comparison of distribution profiles obtained using the nonlinear ODE (7.1) and the standard power-flow equation. This comparison is shown in Fig. 7.7 for the same distribution feeder described in Sect. 7.3. The power-flow

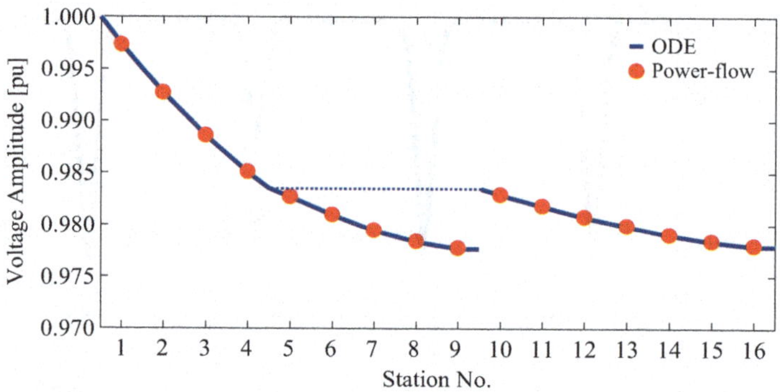

Fig. 7.7 Comparison of distribution voltage profile simulations based on the nonlinear ODE (7.1) and the standard power-flow equation. The two simulations are consistent with the distribution feeder and charging station settings in Fig. 7.3. ©IEEE

equation used here is derived for all charging stations. The *red* points represent the voltage amplitude obtained using the power-flow equation, and the *blue* lines represent values obtained using the nonlinear ODE. The two amplitudes are clearly consistent and suggest that the nonlinear ODE has the same quantitative feature for the current settings of distribution feeder and charging stations.

7.4.2 An Analytical Treatment

Lastly, to exploit the nonlinear ODE for design issues such as AS, we derived an analytical expression of the distribution voltage profile using a partial linearization.

The simulations in Sect. 7.3 suggest that voltage amplitude $v(x)$ does not exhibit a large change in x and stays close to unity (i.e., the nominal value). Thus, it would be relevant to assume that all v values on the right-hand side of (7.1) are equal to unity, which would give us

$$\left.\begin{aligned} \frac{d\theta}{dx} &= -s, \\ \frac{dv}{dx} &= w, \\ \frac{ds}{dx} &= \frac{B \cdot p(x) - G \cdot q(x)}{G^2 + B^2}, \\ \frac{dw}{dx} &= s^2 - \frac{G \cdot p(x) + B \cdot q(x)}{G^2 + B^2}. \end{aligned}\right\} \tag{7.3}$$

The differential equations are linear and can be solved for given $p(x)$ and $q(x)$ values.

Now, as the simplest situation of the non-loading distribution feeder in Fig. 7.1, we suppose that a single station at $x = \xi$ is under an EV charging operation with a constant power of P_{EVs}. It is also supposed that no reactive-power is supplied or consumed at this station, namely $q(\xi) = 0$. Thus, by using Dirac's delta function in $p(x) = p_{\mathrm{EVs}}(x) = P_{\mathrm{EVs}}\delta(x-\xi)$, the following analytical expression of $v(x)$ in (7.3) can be derived:

$$v(x) = \begin{cases} \dfrac{B^2 P_{\mathrm{EVs}}^2}{2Z^4}\left\{\xi^2 - (x-\xi)^2\right\} + \dfrac{G P_{\mathrm{EVs}}}{Z^2}x + 1, \\ \qquad\qquad\qquad\qquad\qquad\qquad 0 \le x \le \xi, \\ \dfrac{B^2 P_{\mathrm{EVs}}^2}{2Z^4}\xi^2 + \dfrac{G P_{\mathrm{EVs}}}{Z^2}\xi + 1, \quad \xi < x \le L, \end{cases}$$
$$:= \mathscr{G}(x,\xi),$$

where $Z := \sqrt{G^2 + B^2}$. Although $\mathscr{G}(x,\xi)$ is not symmetric in this case because of the two-boundary condition problem given as $v(x)|_{x=0}$ and $dv(x)/dx|_{x=L}$, the distribution voltage profile driven by a single station (i.e., EV) can be approximately represented using Green's function [26]. The expression clearly implies that the voltage gradient, i.e., the spatial derivative of $v(x)$, exhibits a discontinuous jump at $x = \xi$. This is consistent with observations from Sect. 7.3. The analytical treatment enables us to quantitatively assess the spatial impact of EV charging and leads to the design of AS considering distribution voltage [21].

7.5 Conclusion

In this chapter, we addressed spatiotemporal impact of charging many EVs on a power distribution grid's voltage profile. Here we introduced a new approach to the assessment problem based on nonlinear ODE and demonstrated it using numerical simulations incorporating operational data from an EV-sharing system demonstration project in Japan. In the companion papers [21, 22], we reported how the nonlinear ODE was exploited for the design of spatial patterns of EV well as charging/discharging operations of EVs for not only provision of AS but also voltage regulation.

Acknowledgements The first author greatly appreciates Mr. Seongcheol Baek and Professor Takashi Hikihara (Kyoto University) for their contributions to the initial research before the work presented here. The work was supported by JST-CREST program (#JP-MJCR15K3).

References

1. P. Fairley, Car sharing could be the EV's killer app. IEEE Spectr. **50**(9), 14–15 (2013)
2. K.K. Tan, K.K.K. Htet, A.S. Narayanan, Mitigation of vehicle distribution in an EV sharing scheme for last mile transportation. IEEE Trans. Intell. Transp. Syst. **16**(5), 2631–2641 (2015)
3. A. Kawashima, N. Makino, S. Inagaki, T. Suzuki, A. Shimizu, Simultaneous optimization of assignment, reallocation and charging of electric vehicles in sharing services, in *Proceeding of IEEE Conference Control Technology and Applications* (2017)
4. Y.G. Rebours, D.S. Kirschen, M. Trotignon, S. Rossignol, A survey of frequency and voltage control ancillary services—Part I: technical features. IEEE Trans. Power Syst. **22**(1), 350–357 (2007)
5. W. Kempton, J. Tomic, Vehicle-to-grid power fundamentals: calculating capacity and net revenue. J. Power Sources **144**(1), 268–279 (2005)
6. J. Tomic, W. Kempton, Using fleets of electric-drive vehicles for grid support. J. Power Sources **168**(2), 459–468 (2007)
7. K. Shimizu, T. Masuta, Y. Ota, A. Yokoyama, Load frequency control in power system using vehicle-to-grid system considering the customer convenience of electric vehicles, in *Proceeding IEEE International Conference Power Systems Technology* (IEEE, Piscataway, 2010), p. 8
8. M.D. Galus, S. Koch, G. Andersson, Provision of load frequency control by PHEVs, controllable loads, and a cogeneration unit. IEEE Trans. Ind. Electron. **58**(10), 4568–4582 (2011)
9. Y. Ota, H. Taniguchi, T. Nakajima, K.M. Liyanage, J. Baba, A. Yokoyama, Autonomous distributed V2G (Vehicle-to-Grid) satisfying scheduled charging. IEEE Trans. Smart Grid **3**(1), 559–564 (2012)
10. C. Goebel, D.S. Callaway, Using ICT-controlled plug-in electric vehicles to supply grid regulation in California at different renewable integration levels. IEEE Trans. Smart Grid **4**(2), 729–740 (2013)
11. K. Clement-Nyns, E. Haesen, J. Driesen, The impact of charging plug-in hybrid electric vehicles on a residential distribution grid. IEEE Trans. Power Syst. **25**(1), 371–380 (2010)
12. K. Clement-Nyns, E. Haesen, J. Driesen, The impact of vehicle-to-grid on the distribution grid. Electr. Power Syst. Res. **81**, 185–192 (2011)
13. OECD/IEA, *Nordic EV Outlook 2018: Insights from Leaders in Electric Mobility* (OECD/IEA, Paris, 2011)
14. M. Chertkov, S. Backhaus, K. Turtisyn, V. Chernyak, V. Lebedev, Voltage collapse and ODE approach to power flows: analysis of a feeder line with static disorder in consumption/production. arXiv:1106.5003 (2011)
15. D. Wang, K. Turtisyn, M. Chertkov, DistFlow ODE: modeling, analyzing, and controlling long distribution feeder, in *Proceeding IEEE 51st Conference on Decision and Control* (CDC) (IEEE, Piscataway, 2012), pp. 5613–5618
16. S. Baek, Y. Susuki, Y. Ota, T. Hikihara, Analysis of distribution voltage profile by ODE model incorporated with power demand/supply data, in *Proceeding of 60th Annual Conference of the Institute of Systems, Control, and Information Engineers* (2016), p. 4 (in Japanese)
17. Y. Susuki, S. Baek, Y. Ota, T. Hikihara, Computer simulation of distribution voltage profile using a nonlinear ODE. IEICE Tech. Rep. **116**(215), 15–20 (2016) , NLP2016-47 (in Japanese)
18. N. Mizuta, Y. Susuki, Y. Ota, A. Ishigame, Simulation of distribution voltage profile by ODE model incorporated with EV sharing data, in *Proceeding of Kansai-Section Joint Convention of Institutes of Electrical Engineering* (2016), p. 2 (in Japanese)
19. R.A. Waraich, M.D. Galus, C. Dobler, M. Balmer, G. Andersson, K.W. Axhausen, Plug-in hybrid electric vehicles and smart grids: investigations based on a microsimulation. Transp. Res. C Emerg. Technol. **28**, 74–86 (2013)

20. J,C. Kelly, T. Ersal, C.-T. Li, B.M. Marshall, S. Kundu, G.A. Keoleian, H. Peng, I.A. Hiskens, J.L. Stein, Sustainability, resiliency, and grid stability of the coupled electricity and transportation infrastructures: case for an integrated analysis. J. Inf. Syst. **21**(4), 04015001 (2015)
21. N. Mizuta, Y. Susuki, Y. Ota, A. Ishigame, An ODE-based design of spatial charging/discharging patterns of in-vehicle batteries for provision of ancillary services, in *Proceeding of IEEE Conference Control Technology and Applications* (2017), 193–198
22. N. Mizuta, Y. Susuki, Y. Ota, A. Ishigame, Synthesis of spatial charging/discharging pattern of in-vehicle batteries for provision of ancillary service and mitigation of voltage impact. IEEE Syst. J. **13**(3), 3443–3453 (2019)
23. Y. Susuki, N. Mizuta, A. Kawashima, Y. Ota, S. Inagaki, T. Suzuki, A continuum approach to assessing the impact of spatio-temporal EV charging to distribution grids, in *Proceeding of IEEE Conference Intelligent Transportation Systems* (2017), pp. 2372–2377
24. J. Machowski, J.W. Bialek, J.R. Bumby, *Power System Dynamics and Stability* (Wiley, Chichester, 1997)
25. M.N. Mojdehi, P. Ghosh, An on-demand compensation function for an EV as a reactive power service provider. IEEE Trans. Veh. Technol. **65**(6), 4572–4583 (2016)
26. G.B. Arfken, H.J. Weber, F.E. Harris, *Mathematical Methods for Physicists: A Comprehensive Guide*, 7th edn. (Elsevier, Amsterdam, 2013)

Part III
Toward Dependable Distributed Energy Management System Using ICT

Chapter 8
Cyber Security for Voltage Control of Distribution Systems Under Data Falsification Attacks

Hideaki Ishii, Shinya Yoshizawa, Yu Fujimoto, Isao Ono, Takashi Onoda, and Yasuhiro Hayashi

8.1 Introduction

Cyber security has become one of the most critical issues facing the safe and reliable operation of energy management systems [9, 17, 31]. Since the supervisory control and data acquisition systems (SCADAs) rely more on communication networks, vulnerability to malicious cyberattacks has grown because more coordination in the overall system is required by the wide use of distributed generation (DG) including photovoltaic (PV) and wind power generation systems in the grid. Cyberattacks on

This work was supported by Japan Science and Technology Agency under the CREST Program, Grant No. JPMJCR15K3 and JPMJCR15K5. Some of the material have appeared in the papers [13] and [24] and are reused with the permissions of the IEEE and the American Automatic Control Council (AACC), respectively.

H. Ishii (✉) · I. Ono
Department of Computer Science, Tokyo Institute of Technology, Yokohama, Japan
e-mail: ishii@c.titech.ac.jp; isao@c.titech.ac.jp

S. Yoshizawa · Y. Fujimoto
Advanced Collaborative Research Organization for Smart Society (ACROSS), Waseda University, Tokyo, Japan
e-mail: s.yoshizawa@aoni.waseda.jp; y.fujimoto@aoni.waseda.jp

T. Onoda
Department of Industrial and Systems Engineering, College of Science and Engineering, Aoyama Gakuin University, Sagamihara-shi, Kanagawa, Japan
e-mail: onoda@ise.aoyama.ac.jp

Y. Hayashi
Department of Electrical Engineering and Bioscience, Waseda University, Tokyo, Japan
e-mail: hayashi@waseda.jp

T. Suzuki et al. (eds.), *Design and Analysis of Distributed Energy Management Systems*, Power Electronics and Power Systems,
https://doi.org/10.1007/978-3-030-33672-1_8

the power infrastructure are physicallydangerous, potentially leading to not only quality degradations, but also damage in facilities and equipment.

Extensive studies on cyber security have been performed for transmission systems in recent years. One important line of research is that on false data injection in state estimation problems. It was found that conventional state estimation methods can become unreliable if the measured data sent from sensors regarding power flow, power generation, loads, and so on are modified. Once such data falsification is introduced in a coordinated fashion into multiple data before they reach the estimator, it is possible that it goes undetected by conventional bad data detection schemes; such attacks are called stealthy in the literature (see e.g., [12, 15, 18, 22, 26]). It is also important to develop security testbeds to carry out more realistic studies [10, 29, 33].

To enhance the security level, robust techniques were introduced in the estimations, and their performance was analyzed [3, 4]. The approach was extended to also take into account data from phasor measurement units (PMUs) whose sampling frequencies are much higher than that of the conventional SCADA measurements [5]. The placement of secure sensors to ensure observability in the presence of attacks was studied as well [32]. On the other hand, a dynamical systems approach for fault detections in the grid has recently been pursued [14, 23].

In this chapter, we focus on a security problem in distribution systems, which have received less attention than in the transmission counterpart (e.g., [20, 27]). In particular, we study voltage regulation, where the objective is to maintain the voltage in the feeders within an admissible range by controlling the secondary voltage of the load ratio control transformer (LRT) at the substation. Such regulation becomes more complicated when a large number of PVs are present due to potential steady state voltage rises [30]. To ease the control, one solution is to deploy sectionizing switches equipped with sensors that send the measurements to the controller over networks [7, 8, 11, 25, 34]. This communication can become subject to cyberattacks in the form of false data injection (FDI), which can indirectly manipulate the controller for the LRT and hence the voltage levels throughout the system.

Our approach for securing voltage regulation is based on two functions. The first is the attack detection algorithm (see [13]). It is a simple rule-based algorithm that considers the normal behavior of the system as well as the feeder line topology. The attacker is assumed to have full knowledge of the control algorithm at the LRT and may attempt to create voltage violation by modifying numerical values of the measurements in a stealthy manner. The detection approach is to monitor if sensor measurements satisfy several simple rules obtained by modeling the normal sensor responses. While the detection algorithm may be effective in finding attacks as well as discouraging attackers to approach the system, it alone cannot guarantee the safe operation of voltage regulation. Therefore, further measures are required to secure the system.

The next step is to develop a control scheme that compensates the detection algorithm in such a way that it enables continuous regulation of the voltage levels. More specifically, the scheme should identify a class of attacks that are effective in disabling tap changes in the LRTs and can induce voltage violation in the system.

After detecting such attacks, the scheme ignores the measurements exhibiting abnormal behavior and replaces them with safer values to be used in the control. The security level is enhanced further as the scheme is shown to successfully deal with situations in which the attacker repeatedly makes additional falsifications. Through detailed simulations based on a model of residential areas in Japan, we demonstrate the effectiveness of the proposed attack detection and resilient control scheme and discuss how the approach can be improved.

Disregarding outliers in data has been adopted as a means to protect systems from faults and cyberattacks in a number of problems. For example, in the context of state estimation [3, 4], robust statistics techniques [21] are used, where measurements resulting in large residue are removed. In the distributed model predictive control framework, issues related to false information provided by some agents have also been discussed [1, 28]. In a more general setting of multi-agent systems, resilient consensus has been studied where each nonfaulty agent leaves out the extreme values from the data sent by neighboring agents in their updates [6, 16]. This type of algorithm can be found in the area of fault-tolerant distributed algorithms in computer science (e.g., [19]).

This chapter is organized as follows. In Sect. 8.2, we provide a brief overview of voltage regulation and the setting of the simulation studies. Section 8.3 describes attacker capability as well as our security approach for the detection algorithm and the resilient control scheme. More details on the attack scenarios for the simulation studies are given in Sect. 8.4. We present the results of the simulations in Sect. 8.5 and then discuss the implications in Sect. 8.6. Concluding remarks are given in Sect. 8.7.

8.2 Voltage Regulation Based on Centralized Control

In this section, we introduce the voltage regulation problem and the centralized control scheme used therein [13].

8.2.1 Centralized Voltage Control

A simple distribution network model is depicted in Fig. 8.1, in which power is distributed in residential and industrial areas through a feeder line from the distribution substation. The objective of voltage regulation is to maintain the voltage levels within a given admissible range to all consumers throughout the areas along the feeder. The admissible range is set approximately 6.6 kV in Japan. The voltage is mainly controlled at the substation and, specifically, by the LRT. The secondary voltage of the LRT depends on the position of its tap. Conventionally, the control of the tap has been taken care by the line drop compensator (LDC), which uses the local measurements of the voltage and current at the substation. Although it is based

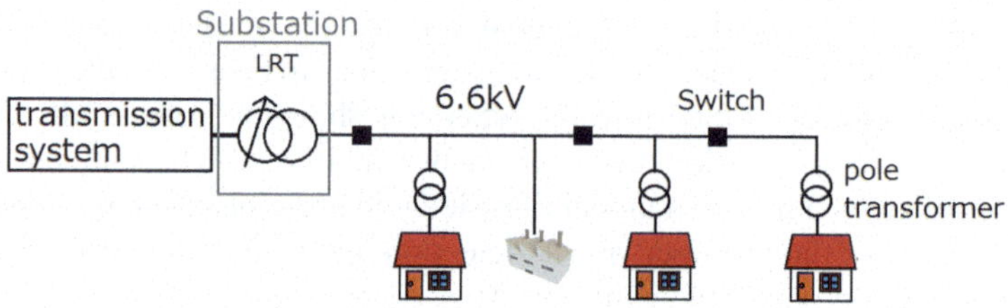

Fig. 8.1 A simple model of a distribution system

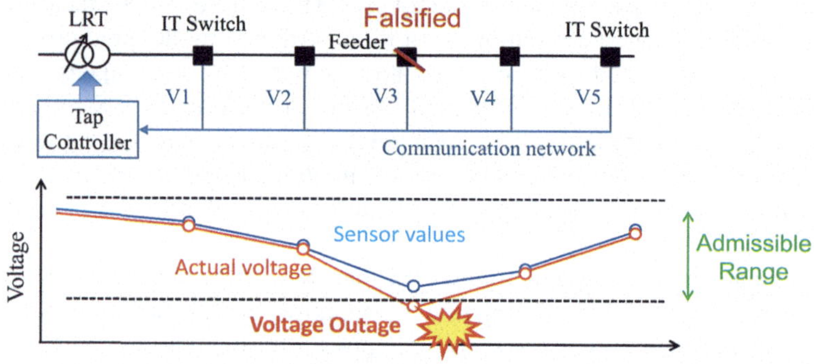

Fig. 8.2 Centralized control for voltage regulation

on the limited information available locally, the LDC approach has been sufficient for voltage regulation over the years.

However, the situation has currently changed as the amount of power generated by DGs has dramatically increased. In the feeder, the voltage profile can vary in a complicated manner as power generated by DGs flows into the distribution system. As the conventional LDC approach does not take account of the voltage at the consumer end, it is not capable of accurately regulating the admissible levels.

One proposed solution [11, 34] is to employ a centralized control scheme by introducing voltage sensors in the sectionizing switches, which are sometimes referred to as IT switches (see Fig. 8.2). These switches transmit their measurements to the substations over a communication network. Then, the real-time voltage profile becomes available to the regulator.

Note that in this centralized control scheme, multiple measurements are collected to control one variable, the secondary voltage of the LRT. It is essentially an integral control to regulate the average value of the voltage measurements around the reference value. We briefly explain its function below. Note that the control approach assumes a radial distribution system, i.e., the feeder lines have no loops.

The controller is a discrete-time system with sampling period T. Let N be the total number of sensors (or nodes). At sampling time kT with integer k, the controller collects the voltage measurement $V_i(k)$ from each node $i \in \{1, \ldots, N\}$. Node i has the admissible range given as

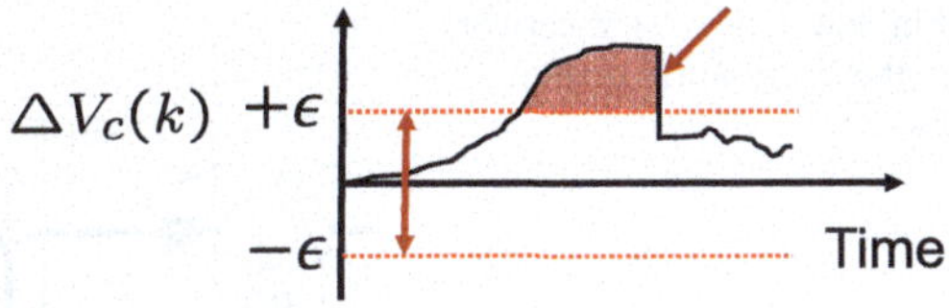

Fig. 8.3 Integration of the voltage deviation index $\Delta V_c(k)$

$$V_{\min,i} \leq V_i(k) \leq V_{\max,i} \text{ for each node } i, \tag{8.1}$$

where $V_{\min,i}$ and $V_{\max,i}$ are the given lower and upper limit values for $V_i(k)$, respectively. Let $V_{\text{ref},i}$ be the reference given by $V_{\text{ref},i} = (V_{\min,i} + V_{\max,i})/2$, and let $\Delta V_i(k)$ be the deviation of $V_i(k)$ from the reference as $\Delta V_i(k) = V_i(k) - V_{\text{ref},i}$. Then, the voltage deviation index $\Delta V_c(k)$ is given by the sum of the minimum and maximum:

$$\Delta V_c(k) = \min_i \Delta V_i(k) + \max_i \Delta V_i(k). \tag{8.2}$$

The portion of this index $\Delta V_c(k)$ that exceeds the deadband determined by the constant $\epsilon > 0$ is integrated in time to yield the tap change index $F(k)$ given as

$$F(k) = \begin{cases} F(k-1) + T \cdot \text{sgn}(\Delta V_c(k)) \cdot (|\Delta V_c(k)| - \epsilon) \\ \qquad\qquad \text{if } |\Delta V_c(k)| > \epsilon, \\ F(k-1) \quad \text{otherwise,} \end{cases} \tag{8.3}$$

where $\text{sgn}(x)$ gives the sign of a scalar x (Fig. 8.3).

The tap position $\text{Tap}(k)$ is switched when the index $F(k)$ becomes larger in magnitude than the threshold F_{ref} as

$$\text{Tap}(k+1) = \begin{cases} \text{Tap}(k) + 1 & \text{if } F(k) < -F_{\text{ref}}, \\ \text{Tap}(k) - 1 & \text{if } F(k) > F_{\text{ref}}, \\ \text{Tap}(k) & \text{otherwise.} \end{cases}$$

Frequent tap changes are to be avoided as they cause wear in the devices.

8.2.2 Simulation Settings for Voltage Regulation

In our study, we use a simple distribution system setup in our simulations [13] which is briefly explained below.

Consider the 6.6 kV distribution network with 14 nodes depicted in Fig. 8.4 that models a small residential district in Japan. Parameters and load profiles from real data are used. The LRT for voltage regulation is connected to a steady voltage source

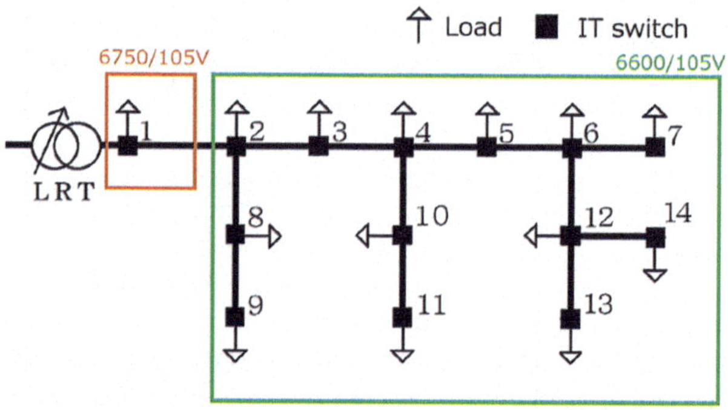

Fig. 8.4 The 6.6 kV network in the distribution system model

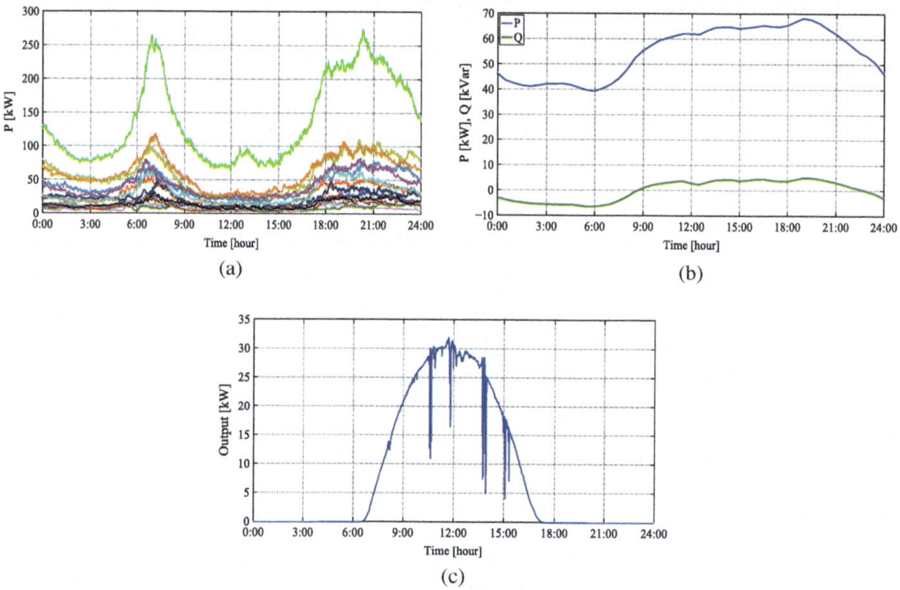

Fig. 8.5 Profiles of data used in simulations (©2016 IEEE. Reprinted with permission from [13]). (**a**) Aggregated low-voltage loads. (**b**) High-voltage loads. (**c**) PV generation

of 66 kV and has the tap width of 30 V. In the figure, each node along the feeder lines represents a sectionizing switch connected to low-voltage 105 V networks via pole transformers. The total contract demand of the system is set as 2113 kVA.

The reference values and the admissible ranges are set as follows: for node 1, $V_{\mathrm{ref},1} = 6750\,\mathrm{V}$ and $[V_{\min,1}, V_{\max,1}] = [6621, 6879]\,\mathrm{V}$; for other nodes, $V_{\mathrm{ref},i} = 6600\,\mathrm{V}$ and $[V_{\min,i}, V_{\max,i}] = [6474, 6726]\,\mathrm{V}$. Node 1 takes different parameters due to the long line between nodes 1 and 2.

All nodes in the system are connected to low-voltage loads in residential areas. Their aggregated power profiles are shown in Fig. 8.5a for a period of 24 h starting at 0:00. Moreover, high-voltage loads are present at some of the nodes, namely, nodes 3, 5, 8, 9, 11, 12, and 13. The profiles taken from an industrial area in Japan are shown in Fig. 8.5b.

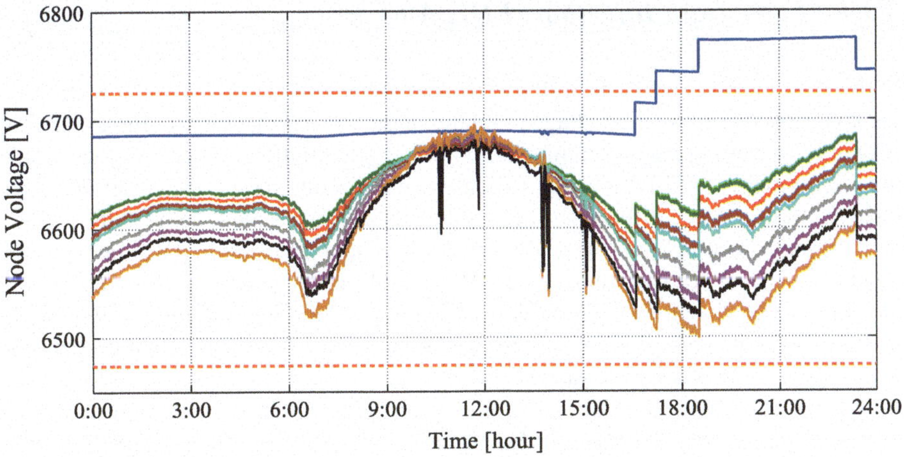

Fig. 8.6 Voltage time responses under normal conditions (©2016 IEEE. Reprinted with permission from [13])

In this model, each residence is equipped with PVs. The generated power is roughly synchronized; we assume that the residential area is within a relatively small district. In the simulation, we use PV profiles from real data taken on a sunny day with some clouds, as shown in Fig. 8.5c.

For this distribution system, the centralized control introduced earlier is capable of regulating the voltage well. The simulation result is presented in Fig. 8.6, where each line indicates the time response of the voltage value of a node over the 24 h periods. The blue piecewise constant line represents the voltage at node 1, whose reference value is larger than that of others; its value is almost the same as that at the LRT. Tap changes occur four times at about 16:30, 17:15, 18:30, and 23:30. The dashed red lines indicate the upper and lower limits of the admissible range for other nodes. In the controller, we set the deadband parameter to be $\epsilon = 90$ V and the threshold to be $F_{\text{ref}} = 15$ V s.

Note that the tap switches upward starting in early evening. This is when the PVs generation slows down, and at the same time, the residents in the district come home and use more power. It is also clear that during the day when PVs generate, voltage values do not follow the order according to the topology in Fig. 8.4.

8.3 Cyberattacks and Security Measures in Voltage Regulation

In this section, we present several scenarios of potential cyberattacks against the voltage regulation system.

8.3.1 False Data Injection (FDI) Attacks

In this centralized control, the voltage measurements of the sensors at the sectionizing switches are sent to the voltage regulator at the substation over a communication network. Therefore, such communication could potentially be attacked. Our focus here is on malicious FDI attacks, which change the raw data transmitted over the channel, thus creating differences between the true measurement value at the sensor and the data received by the LRT. Such false data can manipulate the voltage regulation, which can be dangerous to the distribution system, especially if the attacker is knowledgeable. We will see that even if one measurement is falsified, it can cause voltage violation; that is, voltage at some nodes exceeds the upper or lower limit of the admissible range.

The question that motivates us in this research is to find (1) how and to which extent we can detect these attacks based on the measurement values sent from the sensors, and (2) if we can safely maintain voltage control even if such attacks may be present. We develop detection and resilient control strategies, which will be examined through simulations in the simple yet realistic system setting shown in Fig. 8.4.

To illustrate how the attacks can be harmful and cause voltage violation, we have developed an interactive game in which the player can act as an attacker generating FDI attacks to the sensor values. The game is based on the 14-node distribution system discussed in Sect. 8.2.2, and a screen shot is shown in Fig. 8.7. The player can click on the nodes and select how the values should be modified from four options given in the gray box at the bottom. The two figures in the right represent the true voltage values (top) and the potentially falsified sensor values (bottom). Observe that the voltage violation takes place at certain nodes.

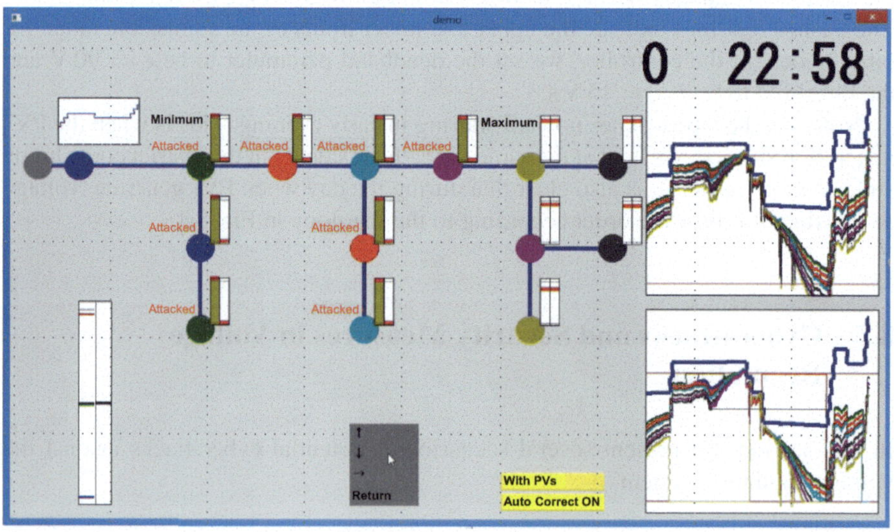

Fig. 8.7 Screen shot of the interactive game for FDI attacks

8.3.2 Detection Algorithm for FDI Attacks

To prevent FDI attacks, we introduced a simple rule-based attack detection algorithm in our previous work [13]. The algorithm monitors the behavior of the sensor values received in real time and alarms whenever it is irregular. We briefly introduce this algorithm. For notation, let $V_i^*(k)$ be the true voltage value at node i at sampling instant kT, where $T > 0$ is the sampling period. Also, let $V_i(k)$ be the corresponding data received by the LRT, which is different from $V_i^*(k)$ when attacked.

The detection algorithm examines the measurement data received from the sensors at every sampling instant. In particular, it checks if the data satisfies several conditions that represent a model of normal measurements. The algorithm consists of four steps as follows:

Step 1 (Voltage values in the admissible range): Find if the measurement value $V_i(k)$ of the voltage at each node i falls within the admissible range as in (8.1).

Step 2 (Order among voltage values): Check if for each node, its voltage value is smaller than those upstream (i.e., closer to the LRT). This step is to be applied only when PVs are not in operation, that is, mainly during the evening and on cloudy days.

Step 3 (Voltage change rates): If tap changes did not occur in the previous time step $(k-1)T$, then check whether $\left|V_i(k) - V_i(k-1)\right| \leq C(k)$ holds, where $C(k)$ represents the upper bound on the rate of changes in $V_i(k)$; this bound may vary according to the level of PV generation.

Step 4 (Lower bound on voltage differences): Check whether $\max_i V_i(k) - \min_i V_i(k) \geq D(k)$ holds, where $D(k)$ is a given (time-varying) lower bound.

Here, in Step 1, the algorithm verifies if voltage violation occurs at any node in the system. Step 2 is applied at hours only when the PVs do not generate power. Then, the power flows from the substation only in the downstream direction along the feeder. In such cases, the voltage becomes a decreasing function of distance from the substation. The remaining two steps simply verify the rates in voltage changes and the differences among the sensor values.

The four steps are illustrated in Fig. 8.8; note that the time responses shown are from a cloudy day with no PV generation during the day. Therefore, Step 2 is applicable throughout the day.

8.3.3 Stealthy Cyberattack Strategies

In reference [13], different attack scenarios were considered. We confirmed that modifying sensor data can manipulate switching in the tap. In particular, some attacks can result in voltage violation in a stealthy manner, i.e., without detection by the algorithm. Under centralized control and the detection algorithm described in Sect. 8.3.2, an effective attack strategy is to modify the measurement values from

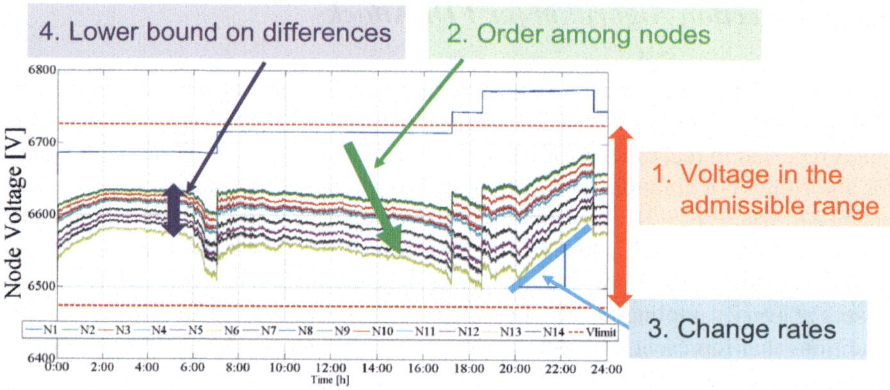

Fig. 8.8 Four rules for attack detection

the sensor so as to disable tap changes at the LRT. This can result in undervoltage, i.e., at times when loads are high, voltage at some nodes may be lower than the admissible limit.

We can determine such an attack strategy by formulating an optimization problem in a simple setting. To proceed, we make the following assumptions:

1. As in Fig. 8.2, the feeder has a line topology with no branching, and the nodes are labeled in ascending order from the substation side.
2. No PV is in operation (e.g., at night or on a cloudy day).
3. The number M of nodes attacked satisfies $M \leq N - 2$.

Regarding the last assumption, node 1 next to the substation is assumed to be difficult to attack as its voltage is easy to estimate at the LRT. We also exclude the case where all nodes would be under attack. Note that because of assumptions 1 and 2, the actual voltage values $V_i^*(k)$ always satisfy the order:

$$V_1^*(k) \geq V_2^*(k) \geq \cdots \geq V_N^*(k). \tag{8.4}$$

We then obtain the following attack strategy:

$$V_i(k) = \begin{cases} \min\{V_1^*(k), V_{\max,2}\} & \text{if } i = 2, \\ V_{N-M+1}^*(k) & \text{if } i = N - M + 2, \ldots, N, \\ V_i^*(k) & \text{otherwise.} \end{cases} \tag{8.5}$$

Here, it is critical to attack node 2 and raise its voltage. It has the highest voltage due to its location and hence will contribute to an increase in $\Delta V_c(k)$ in (8.2) via the term $\max_i \Delta V_i(k)$. Under the given constraints, this attack disables any upward tap changes because they can make the voltage of node 2 higher than the admissible upper limit. The remaining $M - 1$ attacks raise the voltage of nodes towards the end of the feeder so that no undervoltage can be detected by Step 1 of the algorithm.

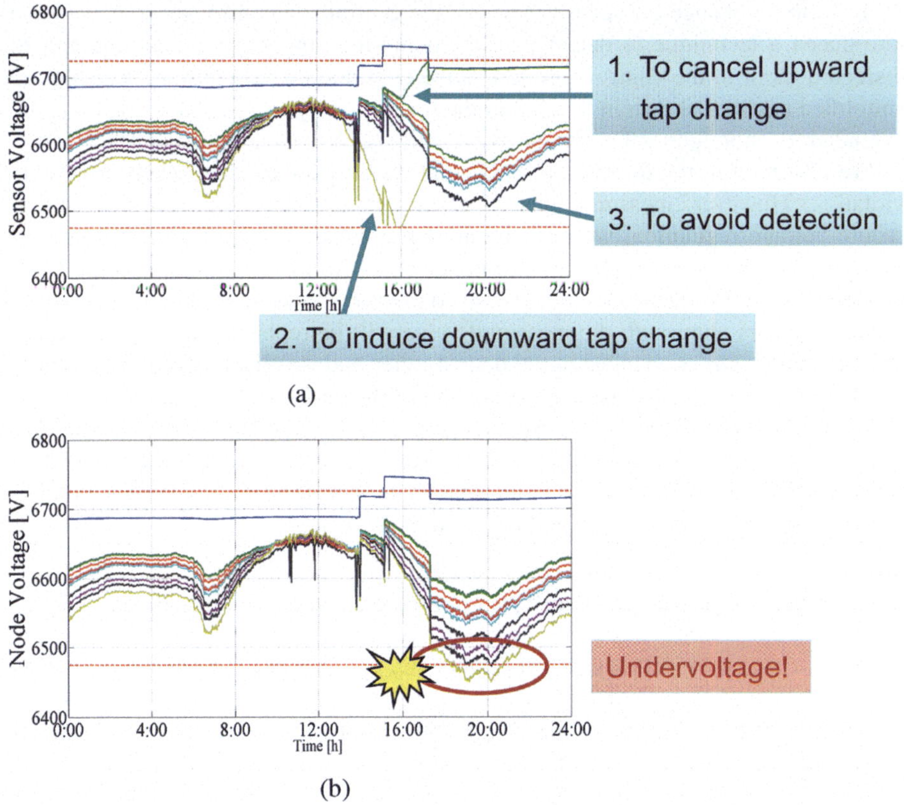

Fig. 8.9 Under the original control with attacks on five nodes. (**a**) Sensor measurements $V_i(k)$. (**b**) Voltage values $V_i^*(k)$

We simulated an instance of such an attack on the distribution system in Fig. 8.4. Here, we generated attacks on five nodes following the attack strategy described above. The results are presented in Fig. 8.9, where the top plot indicates the measured values $V_i(k)$ of the 14 nodes, and the bottom plot shows the true voltage values $V_i^*(k)$. We observe that around 15:00, the node 2 sensor value starts to increase. Moreover, after 17:00, several lines having lower voltage values are raised so that they are away from the lower limit of the admissible range. Indeed, this attack is stealthy in the sense that it stays undetected by the algorithm.

8.3.4 Further Measures for Resilient Control

In voltage regulation, it is important to guarantee safe and continuous operation even in the presence of faults and cyberattacks. Therefore, attack detection provides only a partial solution. After detecting attacks, other measures must be applied to continue the regulation.

To further enhance the security level of the centralized control scheme, we have introduced a technique to detect measurements that appear to be outliers and to disregard them in the control (see [24]). We continue our discussion by using the simplified network shown in Fig. 8.2 in the line topology, but the basic ideas can be extended for more general networks as we will see in the simulations later.

We observed in the discussion so far that stealthy attacks can result in voltage violation. However, measurements of certain nodes play critical roles, and to secure voltage regulation, we must monitor the behavior of those measurements. In particular, the most critical are the nodes closer to the substation and the LRT. As shown in (8.5), the attack strategy was to raise the value at node 2. Whenever its voltage appears abnormal, the measured value will be replaced with a safer value. We can make similar changes in the data at other nodes. When there is a sufficient number of measurements taken, ignoring part of the data should not harm operation. We will more formally state this argument and confirm it in the simulation studies.

We now outline the security-enhanced resilient control scheme presented in reference [24]. Denote by $L \geq 0$ the integer representing the maximum number of measurements to be ignored and replaced in the scheme.

Step 1: At time k, initialize by setting $i = 2$.

Step 2: Check if the measurement $V_i(k)$ of node i satisfies the inequalities

$$V_{i+1}(k) \leq V_i(k) \leq \min\left\{V_1^*(k), V_{\max,i}, V_{i+1}(k) + \delta\right\}, \tag{8.6}$$

where $\delta > 0$ is a threshold (see Fig. 8.10). If not, set $m = i$, and go to Step 3. Otherwise, set $i = i + 1$, and repeat this step. If $i = L + 1$, go to Step 4.

Step 3: Replace the measurement values of nodes $2, \ldots, m$ with that of node $m + 1$ as

$$V_2(k) = V_{m+1}(k), \ldots, V_m(k) = V_{m+1}(k).$$

Step 4: Use the current/updated measurements $V_j(k)$ from all nodes j to obtain $\Delta V_c(k)$ in (8.2), and proceed with the centralized control scheme for time k.

Step 5: Set $k = k + 1$. Go to Step 2.

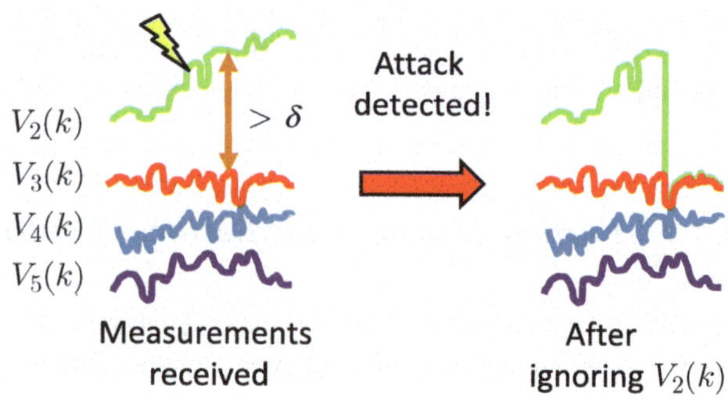

Fig. 8.10 Detection/replacement of false data in the voltage $V_2(k)$ of node 2

The condition in (8.6) is to check if node i satisfies Steps 1 and 2 of the detection algorithm presented in Sect. 8.2. Moreover, it checks if each value is not separated from its immediate neighbor downstream by more than the threshold δ.

8.4 Simulation Settings for Cyber Security

Based on the general discussions regarding the FDI attacks and the security measures, we describe the more specific strategies of the attacker and the system operator for the 14-node distribution system in Fig. 8.4. These strategies will be used in the simulation studies presented in Sect. 8.5.

8.4.1 Attacker Strategy

In the simulation study, we consider cyberattack scenarios where the attacker intends to keep the tap of the LRT unchanged by disabling its control. We observe in simulated results shown in Fig. 8.6 that under normal conditions, four tap switches occurred in the evening. Depending on the attack strategies, some or all of the tap changes can be disabled. This will clearly create problems as the voltage will not properly increase to meet the power demand in real time.

We assume the attacker knows the mechanism of the security measures in the controller. Therefore, the attacker attempts to generate attacks without violating the rules in the detection algorithm, which means that in the evening, the attacks must respect Step 2 of the algorithm in Sect. 8.2 concerning the order among the voltage values.

For the network topology in Fig. 8.4, the voltage order in the evening hours are as follows:

1. $V_1(k) > V_2(k) > \cdots > V_7(k)$,
2. $V_2(k) > V_8(k) > V_9(k)$,
3. $V_4(k) > V_{10}(k) > V_{11}(k)$,
4. $V_6(k) > V_{12}(k) > V_{13}(k)$,
5. $V_{12}(k) > V_{14}(k)$.

In the 14-node distribution system model, the strategy of the attacker introduced in Sect. 8.3.3 can be modified in the following manner:

(A1) The attacker falsifies the sensor data $V_2(k)$ from node 2 and makes it equal to that of node 1, i.e.,

$$V_2(k) = \min\left\{V_1^*(k),\, V_{\max,2}\right\}. \tag{8.7}$$

This attack can disable the upward tap change at the LRT to increase the level of voltage output. In the simulation, to avoid detection of the attack, $V_2(k)$ will be modified smoothly and will gradually reach the condition in (8.7).

(A2) If the attacker sees that the falsification on $V_2(k)$ did not affect the voltage regulation and the number L of attacks satisfies $L \geq 2$, then the sensor value of node 3 can be falsified as

$$V_3(k) = \min\left\{V_1^*(k), V_{\max,3}\right\}. \tag{8.8}$$

This attack can also be generated simultaneously with the one on node 2 in *(A1)*; such an attack increases the chance of influencing the voltage regulation. In either case, it will not violate the voltage order in Step 2 (Sect. 8.3.4) of the algorithm. Note that in view of the network in Fig. 8.4, the attacker can attack node 8 instead of node 3 in the same manner.

(A3) If the number L of attacks satisfies $L \geq 3$, then the attacker may similarly attack either node 4 or 8, depending on the attacks made so far. The attacks can continue with more nodes in the downstream direction of the feeder.

(A4) If attacks made so far are fewer than L, then the attacker can increase the measurement values of nodes whose true voltage may eventually violate the lower limit as $V_i^*(k) < V_{i,\text{ref}}$. This attack is critical since any undervoltage could be overlooked by Step 1 of the detection algorithm. The falsified values should respect the voltage order in Step 2 of the detection algorithm. Accordingly, the attack may make, e.g., $V_{13}(k) = V_{12}^*(k)$ and/or $V_{14}(k) = V_{12}^*(k)$.

8.4.2 Resilient Control

The secure control scheme monitors the behavior of voltage measurements and responds to attacks in the above scenario as follows:

(C1) The controller initially checks if the condition in (8.6) is satisfied for the voltage $V_2(k)$ of node 2. When node 2 is attacked such that the condition in (8.7) are met, eventually the condition in (8.6) at time k will be violated by becoming larger than the threshold as $V_2(k) > V_3(k)+\delta$. Here, parameter δ is set as $\delta = 30$ V, which is equal to the tap width. After detection at this point, the controller sets the value of $V_2(k)$ equal to that of node 3 as

$$V_2(k) = V_3(k).$$

(C2) After detecting the abnormal behavior of node 2 in (C1), the control scheme monitors the behavior of node 3 and node 8, which are the likely targets for the attacker who further tries to disable tap switches. In the event where violation of the condition in (8.6) is detected for either $V_3(k)$ or $V_8(k)$, the value of that node will be set as

$$V_3(k) = V_4(k) \text{ or } V_8(k) = V_9(k).$$

(C3) The detection and replacement of the voltage values will continue until the maximum number L of nodes is detected. Note that due to the branching in the network topology in Fig. 8.4, as the number of attacks increases, the number of nodes that must be monitored will increase.

8.5 Verification Via Simulations

We studied the relationship between the number M of sensors to be falsified/attacked by the attacker and the number L of measurements to be ignored/replaced by the controller through simulations. Here, we discuss how the effectiveness of the resilient control scheme relies on the relationship between the number of attacks and that of values to be ignored.

Case 1 When the number of attacks M is smaller than or equal to the number of measurements L to be ignored, the attacker cannot manipulate the controller's response to suppress its tap switching behavior. The results for the case with $M = L = 3$ are shown in Fig. 8.11, where the top plot (a) shows the measured values $V_i(k)$ received at the controller, and the bottom plot (b) depicts the true voltage values $V_i^*(k)$ of the 14 nodes. Starting around 15:00, the measurements from nodes 2, 3, and 4 were modified to values closer to the upper limit of the admissible range. This modification was detected, and the manipulated values were ignored. This is shown in Fig. 8.11a, where the node values going upward turned to dashed lines around 16:00. Three tap changes took place, which was sufficient and did not result in any voltage violation.

We may conclude that as long as the system is prepared and the upper bound on the number of attacked nodes is known, the system may be operated safely via this control scheme. This attack strategy is similar to the one in Fig. 8.9. Thus, under the original centralized control without the resilient scheme, the system would experience undervoltage.

Case 2 When the number of attacked measurements M is greater than L, the attacker has the advantage and can induce voltage violation.

However, in this case, the voltage violation is visible by observing the measurements, and thus Step 1 of the detection algorithm can find it. Therefore, such attacks are detectable and are less serious as threats.

This case can already occur when the attacker increases the attack by one node and uses $M = 4$ while the controller continues to take $L = 3$. The simulation results are shown in Fig. 8.12. The attack strategy is similar, and sensor values of nodes 2, 3, 4, and 5 are falsified. Attacks are generated around 15:00, but the

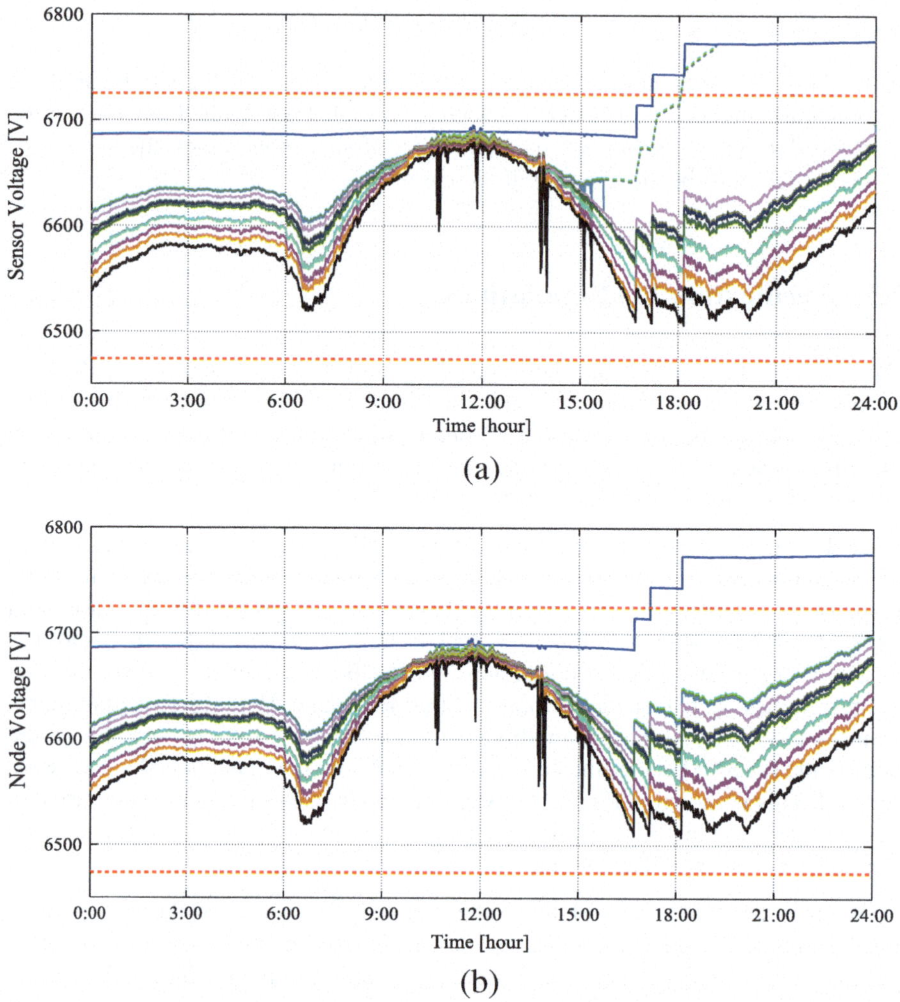

Fig. 8.11 Under resilient control with three attacks (Case 1 with $L = 3$) (©2019 AACC. Reprinted with permission from [24]). (**a**) Sensor measurements $V_i(k)$. (**b**) Voltage values $V_i^*(k)$

controller cannot detect them and the tap switches only once. This is not sufficient to maintain all nodes in the admissible range, and undervoltage occurs at nodes 13 and 14 after 18:00.

In this case, however, the attacks are detectable in the sense that the voltage violation is seen by the controller, and hence, other measures may be taken. However, such detectable attacks are limited to situations when M is relatively small.

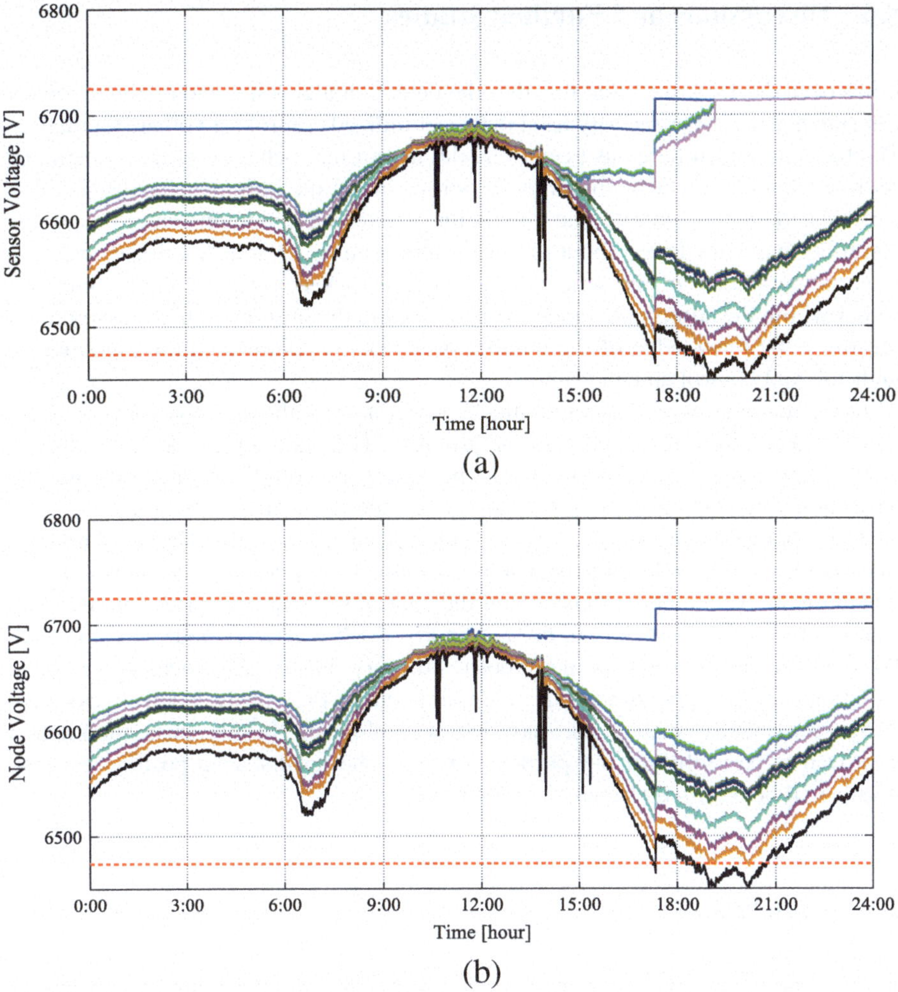

Fig. 8.12 Under resilient control with four attacks (Case 2 with $L = 3$) (©2019 AACC. Reprinted with permission from [24]). (**a**) Sensor measurements $V_i(k)$. (**b**) Voltage values $V_i^*(k)$

If the attacker manages to modify the sensor values of all nodes that are undervoltage, then the attack becomes stealthy, i.e., the detection algorithm cannot spot the attack, at least not from observing the measurements alone. In the current simulation settings, the maximum number of nodes having voltage dip under the lower limit is three. Then, at least four measurements must be falsified to make the attack stealthy.

8.6 Discussions and Further Studies

It has been demonstrated that together the detection algorithm and resilient control can enhance the cyber security level in the centralized control for voltage regulation. The basic approach here has been to model the normal behavior of the system and set rules that can identify abnormal responses in the measurements and reject that particular data for control. The simulation studies have demonstrated that if the controller has knowledge about the maximum number of attacked nodes and uses it as the parameter L, the scheme may succeed in keeping the voltage along the feeder within the admissible range. As discussed in the introduction, the assumption regarding the knowledge of the number of attacks is common in the literature on fault-tolerant control and algorithms.

In this chapter, we considered attacks motivated by those in the form of (8.5), which is a solution to an optimization problem. The main attack idea is to disable upward tap changes by falsifying some measurements to indicate high voltage. We have shown that our secure control scheme is effective against such attacks.

To further enhance security, we are currently considering other classes of attacks. In particular, we have developed an automatic attack generator and carried out large-scale tests to examine the robustness of the control scheme in a systematic way [2]. One instance of the attacks can be seen in Fig. 8.13, where six nodes are attacked. We confirm that the attacks may appear arbitrary, but in fact, there is a voltage violation over the upper limit of the admissible range. Previously, when attacks were designed manually, such attacks have not been obtained. In this sense, the method is proven to be useful for our purpose. Overall, the results seem positive, and the details will be reported elsewhere in the future.

8.7 Conclusion

In this chapter, we have presented our recent studies on enhancing cyber security for the voltage regulation in distributed systems. Security issues have become relevant in such systems due to the increase in DGs, which requires the use of communication channels for collecting more global information. We have provided an overview of the attack detection algorithm from reference [13] and the resilient control scheme from reference [24]. They can cooperatively mitigate the influence of the potential FDI attacks on the voltage measurements. We demonstrated through simulation studies that the system can tolerate attacks while continuing to regulate voltage.

In future research, we will study more complicated attacks based on our recently developed attack generation system and upgrade the resilient control scheme. We will also examine our approach on a hardware simulator of distribution systems.

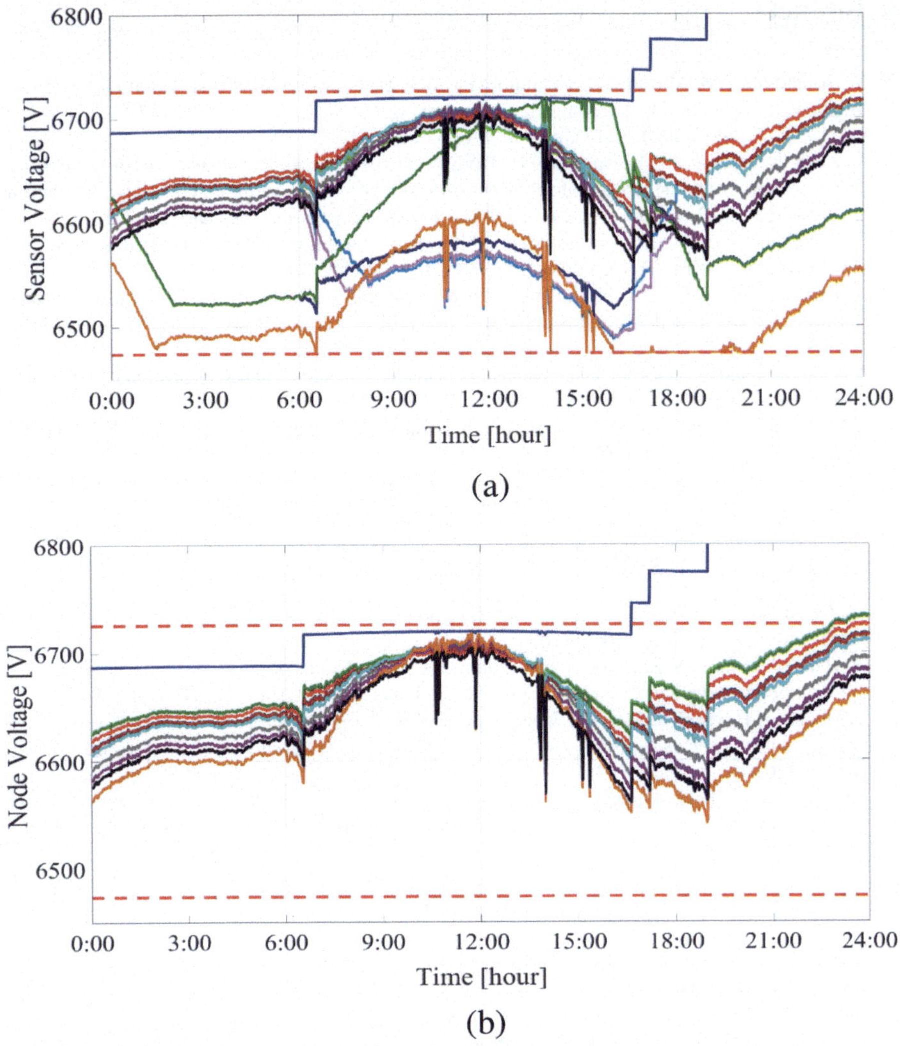

Fig. 8.13 Random attacks on the original control. (**a**) Sensor measurements $V_i(k)$. (**b**) Voltage values $V_i^*(k)$

Acknowledgements The authors are grateful to Yusuke Aoki, Yasunori Isozaki, Mizuki Onogawa, and Atsushi Suzuki for their help in the course of this research.

References

1. W. Ananduta, J.M. Maestre, C. Ocampo-Martinez, H. Ishii, Resilient distributed energy management for systems of interconnected microgrids, in *Proceedings of 57th IEEE Conference on Decision and Control* (2018), pp. 6159–6164
2. Y. Aoki, S. Yoshizawa, Y. Fujimoto, H. Ishii, I. Ono, T. Onoda, Y. Hayashi, A study of cyber attack scenarios against large-scale distribution systems based on centralized voltage regulation (in Japanese), in *Proceedings of SICE Symposium on Systems and Information* (2017)
3. Y. Chakhchoukh, H. Ishii, Coordinated cyber-attacks on the measurement function in hybrid state estimation. IEEE Trans. Power Syst. **30**, 2487–2497 (2015)
4. Y. Chakhchoukh, H. Ishii, Enhancing robustness to cyber-attacks in power systems through multiple least trimmed squares state estimations. IEEE Trans. Power Syst. **31**, 4395–4405 (2016)
5. Y. Chakhchoukh, V. Vittal, G.T. Heydt, H. Ishii, LTS-based robust hybrid SE integrating correlation. IEEE Trans. Power Syst. **32**, 3127–3135 (2017)
6. S.M. Dibaji, H. Ishii, R. Tempo, Resilient randomized quantized consensus. IEEE Trans. Autom. Control **63**, 2508–2522 (2018)
7. M.E. Elkhatib, R. El-Shatshat, M.M.A. Salama, Novel coordinated voltage control for smart distribution networks with DG. IEEE Trans. Smart Grid **2**, 598–605 (2011)
8. H.E. Farag, E.F. El-Saadany, R. Seethapathy, A two ways communication-based distributed control for voltage regulation in smart distribution feeders. IEEE Trans. Smart Grid **3**, 271–281 (2012)
9. M. Govindarasu, P.W. Bauer (guest editors), Special section on keeping the smart grid safe. IEEE Power Energy Mag. **10**(1) (2012)
10. A. Hahn, A. Ashok, S. Sridhar, M. Govindarasu, Cyber-physical security testbeds: architecture, application, and evaluation for smart grid. IEEE Trans. Smart Grid **4**(2), 847–855 (2013)
11. Y. Hanai, Y. Hayashi, J. Matsuki, Voltage control for loop distribution system with renewable energy sources, in *Proceedings of International Conference on Renewable Energies and Power Quality* (2010)
12. J.M. Hendrickx, K.H. Johansson, R.M. Jungers, H. Sandberg, K.C. Sou, Efficient computations of a security index for false data attacks in power networks. IEEE Trans. Autom. Control **59**, 3194–3208 (2014)
13. Y. Isozaki, S. Yoshizawa, Y. Fujimoto, H. Ishii, I. Ono, T. Onoda, Y. Hayashi, Detection of cyber attacks against voltage control in distribution power grids with PVs. IEEE Trans. Smart Grid **7**, 1824–1835 (2016)
14. O. Kosut, Malicious data attacks against dynamic state estimation in the presence of random noise, in *Proceedings of IEEE Global Conference on Signal and Information Processing* (2013), pp. 261–264
15. O. Kosut, L. Jia, R. Thomas, L. Tong, Malicious data attacks on the smart grid. IEEE Trans. Smart Grid **2**, 645–658 (2011)
16. H.J. LeBlanc, H. Zhang, X. Koutsoukos, S. Sundaram, Resilient asymptotic consensus in robust networks. IEEE J. Sel. Areas Commun. **31**, 766–781 (2013)
17. T.G. Lewis, *Critical Infrastructure Protection in Homeland Security: Defending a Networked Nation*, 2nd edn. (Wiley, Hoboken, 2015)
18. Y. Liu, P. Ning, M. Reiter, False data injection attacks against state estimation in electric power grids. ACM Trans. Inf. Syst. Secur. **14**, 1–33 (2011)
19. N.A. Lynch, *Distributed Algorithms* (Morgan Kaufmann, Burlington, 1996)
20. M. Ma, A. Teixeira, J. van den Berg, P. Palensky, Voltage control in distributed generation under measurement falsification attacks, in *Proceedings of 20th IFAC World Congress* (2017), pp. 8379–8384
21. R.A. Maronna, R.D. Martin, V.J. Yohai, *Robust Statistics: Theory and Methods* (Wiley, Hoboken, 2006)

22. Y. Mo, T.H. Kim, K. Brancik, D. Dickinson, H. Lee, A. Perrig, B. Sinopoli, Cyber-physical security of a smart grid infrastructure. Proc. IEEE **100**(1), 195–209 (2012)
23. H. Nishino, H. Ishii, Distributed detection of cyber attacks and faults for power systems, in *Proceedings of 18th IFAC World Congress* (2014), pp. 11932–11937
24. M. Onogawa, S. Yoshizawa, Y. Fujimoto, H. Ishii, I. Ono, T. Onoda, Y. Hayashi, Enhancing security for voltage control of distribution systems under data falsification attacks, in *Proceedings of American Control Conference* (2019)
25. T. Senjyu, Y. Miyazato, A. Yona, N. Urasaki, T. Funabashi, Optimal distribution voltage control and coordination with distributed generation. IEEE Trans. Power Delivery **23**, 1236–1242 (2008)
26. K. Sou, H. Sandberg, K.H. Johansson, Electric power network security analysis via minimum cut relaxation, in *Proceedings of 50th IEEE Conference on Decision and Control* (2011), pp. 4054–4059
27. A. Teixeira, G. Dán, H. Sandberg, R. Berthier, R. Bobba, A. Valdes, Security of smart distribution grids: data integrity attacks on integrated Volt/VAR control and countermeasures, in *Proceedings of American Control Conference* (2014), pp. 4372–4378
28. P. Velarde, J.M. Maestre, H. Ishii, R.R.Negenborn, Vulnerabilities in Lagrange-based distributed model predictive control. Optimal Control Appl. Methods **39**, 601–621 (2018)
29. V. Venkataramanan, A. Srivastava, A. Hahn, Real-time co-simulation testbed for microgrid cyber-physical analysis, in *Proceedings of Workshop on Modeling and Simulation of Cyber-Physical Energy Systems (MSCPES)* (2016), pp. 1–6
30. R.A. Walling, R. Saint, R.C. Dugan, J. Burke, Summary of distributed resources impact on power delivery systems. IEEE Trans. Power Delivery **23**, 1636–1644 (2008)
31. W. Wang, Z. Lu, Cyber security in the smart grid: survey and challenges. Comput. Netw. **57**(5), 1344–1371 (2013)
32. I. Watanabe, K. Masutomi, I. Ono, Robust meter placement against false data injection attacks on power system state estimation, in *Proceedings of ICONIP*. Lecture Notes in Computer Science, vol. 8226 (Springer, Berlin, 2013), pp. 569–576
33. T. Yardley, R. Berthier, D. Nicol, W.H. Sanders, Smart grid protocol testing through cyber-physical testbeds, in *Proceedings of IEEE PES Innovative Smart Grid Technologies Conference (ISGT)* (2013), pp. 1–6
34. S. Yoshizawa, Y. Hayashi, M. Tsuji, E. Kamiya, Centralized voltage control method of load ratio control transformer and step voltage regulator for bank fault restoration, in *Proceedings of 3rd IEEE PES Innovative Smart Grid Technologies Europe* (2012)

Chapter 9
Machine Learning Based Intrusion Detection in Control System Communication

Takashi Onoda

9.1 Current Status for Intrusion Detection in Control System

The importance of cybersecurity has increased with the networked and highly complex structure of computer systems, and the increased value of information. Securing the network perimeter of a system is no longer considered sufficient to secure the system, and many layers of security, or defense in depth [1] is necessary.

Intrusion detection is one layer of security for a system, and is defined as "the act of detecting actions that attempt to compromise the confidentiality, integrity or availability of a resource" [2]. A basic intrusion detection in general checks IP (Internet Protocol) addresses, communication ports, and protocols at the TCP/UDP (Transmission Control Protocol/User Datagram Protocol) level for non-legitimate connections. Constructing intrusion detection rules to match attacks does not necessarily require knowledge of the protocol details of the applications being used. Such rules are necessary to detect many attacks, and are proved to be highly effective [3, 4].

We believe that it is effective to consider the feature representation of the network traffic in control system communications without sequence patterns. And also we believe that it is effective to consider the sequence characteristics and the feature representation of the network traffic in control system communications with the sequence patterns.

Manually creating all the intrusion detection rules becomes impractical in a large system, and there has been work using machine learning for the creation of detection rules [5]. If there have been almost no observed cyberattacks, and very

T. Onoda (✉)
Department of Industrial and Systems Engineering, College of Science and Engineering,
Aoyama Gakuin University, Sagamihara-shi, Kanagawa, Japan
e-mail: onoda@ise.aoyama.ac.jp

T. Suzuki et al. (eds.), *Design and Analysis of Distributed Energy Management Systems*, Power Electronics and Power Systems,
https://doi.org/10.1007/978-3-030-33672-1_9

little attack data is available, it is difficult to construct a binary class classification problem for machine learning, which uses both the legitimate communication data and the attack data. In such cases, anomaly detection for intrusion detection would be a good choice [6]. In this chapter, we adopt local outlier factor (LOF), one-class support vector machine (OCSVM), and support vector domain description (SVDD) as anomaly detection methods for the control system communication data without the sequence patterns. And we compare these anomaly detection methods with C4.5 and support vector machine (SVM) in our experiments.

There are many algorithms that can reflect sequence characteristics for the control system communication data with the sequence patterns. The sequence data of the network traffic caused by the control system are probabilistic. In such a case, a hidden Markov model (HMM) and a conditional random field (CRF) can reflect sequence characteristics. In this chapter, we compare these two probabilistic models that reflect the sequence characteristics of network traffic into the intrusion detection system with one-class support vector machine (one-class SVM), support vector data description (SVDD), and rule based intrusion detection method [7] in our experiments. The learning only utilizes normal data, assuming that there is no prior knowledge on attacks in the system. Section 9.2 describes about the related works. Section 9.3 explains an intrusion detection for the control system communication data without the sequence patterns and briefly describes two binary classification methods such as C4.5 and SVM, and three anomaly detection methods such as LOF, OCSVM, and SVDD. Section 9.4 shows the results of the experiments to compare two binary classification methods with three anomaly detection methods. Section 9.5 explains an intrusion detection considering sequences of control communication with the sequence patterns and briefly describes two probabilistic models such as HMM and CRF. Section 9.6 shows the results of the experiments to compare two probabilistic models with OCSVM, SVDD, and rule based intrusion detection method. Sections 9.5 and 9.6 are the content described in the reference [8, 29]. Finally, we conclude our research.

9.2 Various Approaches

Many machine learning techniques have been applied to intrusion detection, including support vector machine (SVM), C4.5, Bayesian network, one-class support vector machine (OCSVM), support vector data description (SVDD), hidden Markov models (HMMs), and conditional random fields (CRF).

The reference [9] proposes an encrusted CRF in intrusion detection systems. The paper shows the performance of the proposed method is better than the performance of naive Bayes and decision tree. And the encrusted CRF can be obtained with high efficiency. The experiments use the benchmark KDD '99 data and real dataset [10]. These datasets are information system communication data. The paper does not deal with the characteristics of control system communication.

The reference [11] makes a comparative study of HMM and SVM in anomaly intrusion detection. This paper aims to analyze the performance of HMM and SVM for anomaly intrusion detection. These techniques discriminate between normal and abnormal behavior of network traffic. The specific focus of this study is to investigate and identify distinguishable TCP services that comprise of both normal and abnormal types of TCP packets. The publicly available KDD Cup 1999 dataset has been used in training and evaluation of such techniques. Experimental results demonstrate that the HMM and SVM make good performance. This research deals with intrusion detection problem as a two-class classification problem and also does not deal with the characteristics of control system communication.

Reference [12] proposes an intrusion detection system based on Bayesian network modeling. The research propose an IDS based on Bayesian network classification modeling technique. The proposed IDS has been evaluated against the KDD DAPRA dataset which was designed for network IDS evaluation. The research methodology consists of four different Bayesian networks as classification models, where each of these classifier models is interconnected and communicated to predict on incoming network traffic data. Each designed Bayesian network model is capable of detecting a major category of attack such as denial of service (DoS). However, all four Bayesian networks work together to pass the information of the classification model to calibrate the IDS system. The proposed IDS shows the ability of detecting novel attacks by continuing learning with different datasets. The testing dataset constructed by sampling the original KDD dataset to contain balance number of attacks and normal connections. The experiments show that the proposed system is effective in detecting attacks in the test dataset and is highly accurate in detecting all major attacks recorded in DARPA dataset. This research also deals with intrusion detection problem as a two-class classification problem and also does not deal with the characteristics of control system communication.

Our research focuses on the networked control systems such as an intelligent distribution network system and so on are appearing and the cybersecurity will become very important for the control systems in the near future. However, we have few actual cyberattacks against the control systems. The intrusion detection should be developed by using only normal control system communication. In the other words, our problem is one-class classification problem and is not two-class classification problem. And we use the characteristics of control system communication as a new feature to detect intrusion.

9.3 Intrusion Detection for Control System Communication Without Sequence Patterns

This section describes the feature representation of control systems' network traffic that is very important for intrusion detection in control system communication without the sequence patterns. And two binary classification methods, which can classify normal network traffic or anomaly network traffic, are briefly explained in this section. In this section, we describe briefly three anomaly detection methods which can be compared with two binary classification methods.

9.3.1 The Feature Representation for Intrusion Detection in Control System Communication

We introduce two control systems which do not have sequence patterns in the control system communication. The two control systems are a water storage tank control system and a gas pipeline control system. These two control systems were introduced by Wei [13].

Water Storage Tank Control System The water storage tank control system models oil storage tanks found in the petrochemical industry. Petrochemical refineries use oil and oil byproducts to produce gasoline, kerosene, diesel, and many types of plastics. In a common configuration, oil arrives by sea and is pumped into oil storage tanks to provide a consistent supply of oil to the refinery operation. Oil storage tank control systems similar to the one modeled by the water storage tank control system are used to monitor oil inventory and distribute oil to refinery processes. Water was substituted for oil for safety reasons.

The water storage tank control system contains primary storage tank and secondary water storage, a pump to move water from the secondary storage tank to the primary storage tank, a gravity fed manual relieve valve which allows water to flow from the primary to secondary storage tank, and a sensor which provides the water level in the primary storage tank as a percentage of total capacity. The secondary storage tank is not a feature of an industrial oil storage tank. The secondary storage tank is used to provide a destination for water when it leaves the primary storage tank and a source for water to fill the primary storage tank. The water storage tank control system is a closed loop.

Gas Pipeline Control System The gas pipeline control system models a gas pipeline used to move natural gas or other petroleum products to market. Cyber penetration of control systems monitoring and controlling a gas pipeline control systems can lead to loss of the visibility and loss of control of the gas pipeline. The gas pipeline control system contains a closed loop gas pipeline connected to an air pump which pumps air into the pipeline. A manual release valve and a solenoid release valve are available to release air pressure from the pipeline. A pressure sensor is attached to the pipeline which allows pressure visibility at the pipeline and remotely on an HMI screen.

The control system communication system of these two control systems does not have the control sequence patterns, which are the order of the controlled equipments. For example, the equipment A has to start after checking the run of the equipment B in a control system. This is an example of the control sequence pattern.

In this case, it is very important for intrusion detection in the control system to generate good feature representation. The communication of these control systems consists of control systems' network traffic information such as IP address, control command, and so on and sensor information such as water level, flow level, and so on. It is hard to detect intrusion by learning just IP address, water level, and so on. In order to detect intrusion effectively, we need to generate the frequency of IP address from the control systems' network traffic information and generate the discrepancy between maximum water level and the present water level and the discrepancy between minimum water level and the present water level from the sensor information. Therefore, it is very important for intrusion detection to generate the good feature representation from control systems' network traffic information and sensor information. When the good feature representation is generated, we can use some machine learning methods. Next, we introduce two binary classification methods and three anomaly detection methods as machine learning methods.

9.3.2 *Binary Classification Methods*

Next, we introduce C4.5 and support vector machine as binary classification methods. These methods are adopted in this experiment.

9.3.2.1 C4.5

C4.5 is an algorithm used to generate a decision tree developed by Ross Quinlan. C4.5 is an extension of Quinlan's earlier ID3 algorithm. The decision trees generated by C4.5 can be used for classification, and for this reason, C4.5 is often referred to as a statistical classifier.

C4.5 builds decision trees from a set of training data in the same way as ID3, using the concept of information entropy. The training data is a set $S = \{s_1, s_2, \ldots\}$ of already classified samples. Each sample s_i consists of a p-dimensional vector $(x_{1,i}, x_{2,i}, \ldots, x_{p,i})$, where the x_j represent feature values or features of the sample, as well as the class in which s_i falls.

At each node of the tree, C4.5 chooses the feature of the data that most effectively splits its set of samples into subsets enriched in one class or the other. The splitting criterion is the normalized information gain. The feature with the highest normalized information gain is chosen to make the decision. The C4.5 algorithm then recurses on the partitioned sublists.

In pseudocode, the C4.5 algorithm for building decision trees is

1. Check for the following base cases:

   (a) All the samples in the list belong to the same class. When this happens, it simply creates a leaf node for the decision tree saying to choose that class.

(b) None of the features provides any information gain. In this case, C4.5 creates a decision node higher up the tree using the expected value of the class.
(c) Instance of previously unseen class encountered. Again, C4.5 creates a decision node higher up the tree using the expected value.

2. For each feature a, find the normalized information gain ratio from splitting on a.
3. Let a_best be the feature with the highest normalized information gain.
4. Create a decision node that splits on a_best.
5. Recur on the sublists obtained by splitting on a_best, and add those nodes as children of node.

9.3.2.2 Support Vector Machine (SVM)

Formally, the support vector machine (SVM) [14] like any other classification method aims to estimate a classification function $f : \mathscr{X} \rightarrow \{\pm 1\}$ using labeled training data from $\mathscr{X} \times \{\pm 1\}$. Moreover this function f should even classify unseen examples correctly.

In order to construct good classifiers by learning, two conditions have to be respected. First, the training data must be an unbiased sample from the same source (pdf) as the unseen test data. This concerns the experimental setup. Second, the size of the class of functions from which we choose our estimate f, the so-called capacity of the learning machine, has to be properly restricted according to statistical learning theory [14]. If the capacity is too small, complex discriminant functions cannot be approximated sufficiently well by any selectable function f in the chosen class of functions—the learning machine is too simple to learn well. On the other hand, if the capacity is too large, the learning machine bears the risk of overfitting. In neural network training, overfitting is avoided by early stopping, regularization, or asymptotic model selection [15–18].

For SV learning machines that implement linear discriminant functions in feature spaces, the capacity limitation corresponds to finding a large margin separation between the two classes. The margin ϱ is the minimal distance of training points $(\boldsymbol{x}_1, y_1), \ldots, (\boldsymbol{x}_i, y_i), \boldsymbol{x}_i \in \boldsymbol{R}, y_i \in \{\pm 1\}$ to the separation surface, i.e.,

$$\rho = \min_{i=1,\ldots,\ell} \rho(z_i, f), \tag{9.1}$$

where $z_i = (\boldsymbol{x}_i, y_i)$ and

$$\rho(z_i, f) = y_i f(\boldsymbol{x}_i), \tag{9.2}$$

and f is the linear discriminant function in some feature space

$$f(\boldsymbol{x}) = (\boldsymbol{w} \cdot \Phi(\boldsymbol{x})) + b = \sum_{i=1}^{\ell} \alpha_i y_i (\Phi(\boldsymbol{x}_i) \cdot \Phi(\boldsymbol{x})) + b, \tag{9.3}$$

with $\boldsymbol{w}$ expressed as $\boldsymbol{w} = \sum_{i=1}^{\ell} \alpha_i y_i \Phi(\boldsymbol{x}_i)$. The quantity Φ denotes the mapping from input space $\mathscr{X}$ by explicitly transforming the data into a feature space $\mathscr{F}$ using $\Phi : \mathscr{X} \rightarrow \mathscr{F}$. In order to train and classify, all that SVMs use are dot products of pairs of data points $\Phi(\boldsymbol{x}), \Phi(\boldsymbol{x}_i) \in \mathscr{F}$ in feature space (cf. Eq. (9.3)). A kernel function k allows to implicitly define the feature space (Mercer's theorem, e.g., [19]) via

$$k(\boldsymbol{x}, \boldsymbol{x}_i) = (\Phi(\boldsymbol{x}) \cdot \Phi(\boldsymbol{x}_i)). \tag{9.4}$$

By using different kernel functions, the SVM algorithm can construct a variety of learning machines, some of which coincide with classical architectures:

Polynomial classifiers of degree d:

$$k(\boldsymbol{x}, \boldsymbol{x}_i) = (\kappa \cdot (\boldsymbol{x} \cdot \boldsymbol{x}_i) + \Theta)^d \tag{9.5}$$

Neural networks (sigmoidal):

$$k(\boldsymbol{x}, \boldsymbol{x}_i) = \tanh(\kappa \cdot (\boldsymbol{x} \cdot \boldsymbol{x}_i) + \Theta) \tag{9.6}$$

Radial basis function classifiers:

$$k(\boldsymbol{x}, \boldsymbol{x}_i) = \exp\left(-\frac{\|\boldsymbol{x} - \boldsymbol{x}_i\|^2}{\sigma}\right) \tag{9.7}$$

Note that there is no need to use or know the form of Φ, because the mapping is never performed explicitly. The introduction of Φ in the explanation above was for purely didactical and not algorithmical purposes. Therefore, we can computationally afford to work in implicitly very large (e.g., 10^{10}-dimensional) feature spaces. SVM can avoid overfitting by controlling the capacity and maximizing the margin. Simultaneously, SVMs learn which of the features implied by the kernel k are distinctive for the two classes, i.e., instead of finding well-suited features by ourselves (which can often be difficult), we can use the SVM to select them from an extremely rich feature space.

With respect to good generalization, it is often profitable to misclassify some outlying training data points in order to achieve a larger margin between the other training points (see Fig. 9.1 for an example).

This soft-margin strategy can also learn non-separable data. The trade-off between margin size and number of misclassified training points is then controlled by the regularization parameter C (softness of the margin). The following quadratic program (QP) (see, e.g., [14, 20]):

$$\begin{aligned} &\min\ \|\boldsymbol{w}\|^2 + C \sum_{i=1}^{\ell} \xi_i \\ &\text{s.t.} \quad \rho(z_i, \boldsymbol{\alpha}) \geq 1 - \xi_i \quad \text{for all } 1 \leq i \leq \ell \\ &\qquad\quad \xi_i, \alpha_i \geq 0 \qquad\qquad \text{for all } 1 \leq i \leq \ell \end{aligned} \tag{9.8}$$

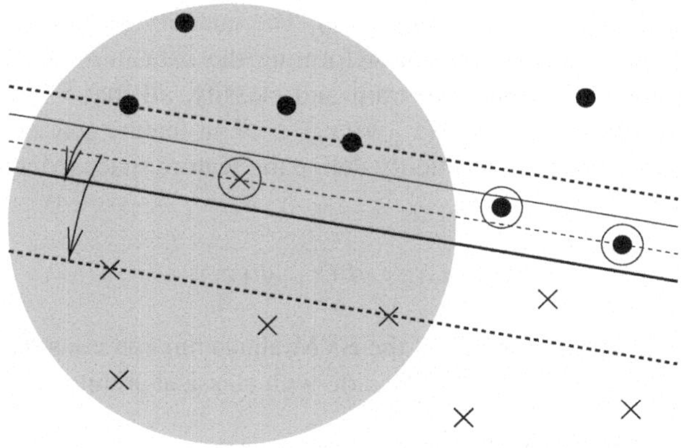

Fig. 9.1 A binary classification toy problem: This problem is to separate black circles from crosses. The shaded region consists of training examples, the other regions of test data. The data can be separated with a margin indicated by the slim dashed lines, implicating the slim solid line as discriminate function. Misclassifying one training example (a circled white circle) leads to a considerable extension (arrows) of the margin (fat dashed and solid lines) and this fat solid line can classify two test examples (circled black circles) correctly

leads to the SV soft-margin solution allowing for some errors.

9.3.3 *Anomaly Detection Methods*

Now, we briefly explain anomaly detection methods which are adopted in this experiment

9.3.3.1 Local Outlier Factor (LOF)

In anomaly detection, the local outlier factor (LOF) is an algorithm for finding anomalous data points by measuring the local deviation of a given data point with respect to its neighbors.

The local outlier factor is based on a concept of a local density, where locality is given by k nearest neighbors, whose distance is used to estimate the density. By comparing the local density of an object to the local densities of its neighbors, one can identify regions of similar density, and points that have a substantially lower density than their neighbors. These are considered to be outliers.

The local density is estimated by the typical distance at which a point can be "reached" from its neighbors. The definition of "reachability distance" used in LOF is an additional measure to produce more stable results within clusters (Figs. 9.2 and 9.3).

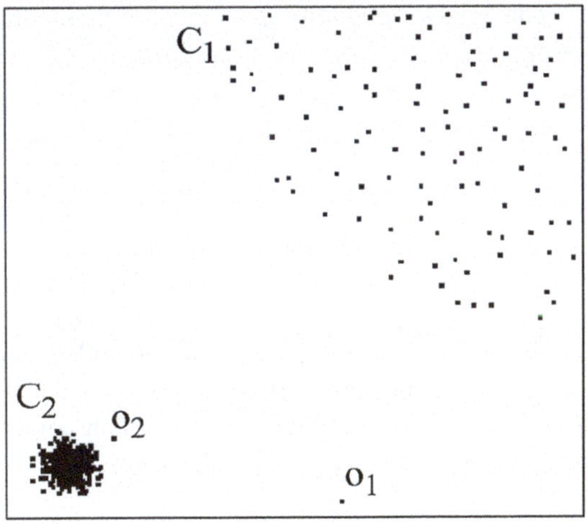

Fig. 9.2 Basic idea of LOF: comparing the local density of a point with the densities of its neighbors. o_1 has a much lower density than points of C_1 and o_2 has a much lower density than points of C_2. So that, o_1 and o_2 are outlier

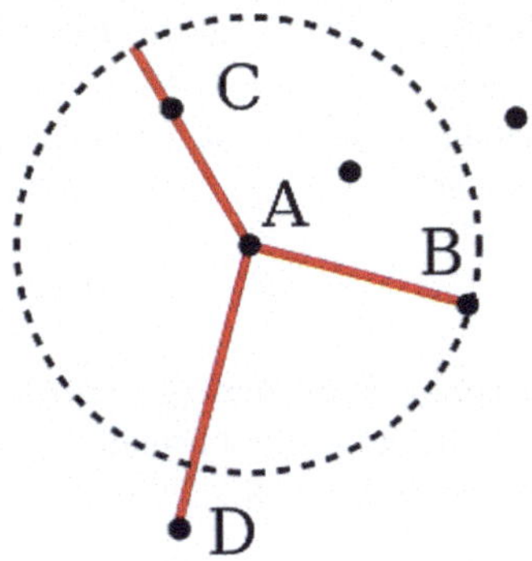

Fig. 9.3 Illustration of the reachability distance. Objects B and C have the same reachability distance ($k = 3$), while D is not a k nearest neighbor

Let k-distance(A) be the distance of the object A to the k-th nearest neighbor. Note that the set of the k nearest neighbors includes all objects at this distance, which can in the case of a "tie" be more than k objects. We denote the set of k nearest neighbors as $N_k(A)$.

This distance is used to define what is called reachability distance: reachability-distance$_k(A, B) = \max\{k\text{-distance}(B), d(A, B)\}$

The local reachability density of an object A is defined by

$$\mathrm{lrd}_k(A) := \left(\frac{|N_k(A)|}{\sum_{B \in N_k(A)} \text{reachability-distance}_k(A, B)} \right) \tag{9.9}$$

which is the inverse of the average reachability distance of the object A from its neighbors. Note that it is not the average reachability of the neighbors from A, but the distance at which A can be "reached" from its neighbors. With duplicate points, this value can become infinite.

The local reachability densities are then compared with those of the neighbors using

$$\mathrm{LOF}_k(A) := \frac{\sum_{B\in N_k(A)} \frac{\mathrm{lrd}(B)}{\mathrm{lrd}(A)}}{|N_k(A)|} \tag{9.10}$$

which is the average local reachability density of the neighbors divided by the object's own local reachability density. A value of approximately 1 indicates that the object is comparable to its neighbors (and thus not an outlier). A value below 1 indicates a denser region (which would be an inlier), while values significantly larger than 1 indicate outliers. $LOF(k) \sim 1$ means similar density as neighbors, $LOF(k) < 1$ means higher density than neighbors (Inlier), $LOF(k) > 1$ means lower density than neighbors (Outlier)

9.3.3.2 One-Class Support Vector Machine (OCSVM)

Here, we adopt the OCSVM proposed by Schölkopf. First, consider the training dataset:

$$D = (\boldsymbol{x}_i, y_i)|\boldsymbol{x}_i \in R^n, y_i = +1_{i=1}^{\ell} \tag{9.11}$$

where $\boldsymbol{x}_i$ is the feature vector with dimension n, $y_i = +1$ means all training patterns.

The algorithm basically separates all the data points from the origin. Suppose the hyperplane has the form:

$$\boldsymbol{w}^T\phi(\boldsymbol{x}) - \rho = 0 \tag{9.12}$$

then the distance from the hyperplane to the origin is $\frac{\rho}{\|\boldsymbol{w}\|}$. Maximizing the distance results to solving the following quadratic programming problem:

$$\begin{array}{lll} \min_{\boldsymbol{w},\xi_i,\rho} \frac{1}{2}\|\boldsymbol{w}\|^2 + \frac{1}{\nu\ell}\sum_{i=1}^{\ell}\xi_i - \rho & & \\ \text{s.t.} & \boldsymbol{w}^T\phi(\boldsymbol{x}_i) \geq \rho - \xi_i l, & \text{for all } 1 \leq i \leq \ell \\ & \xi \geq 0, & \text{for all } 1 \leq i \leq \ell \end{array} \tag{9.13}$$

Here, $\phi(\boldsymbol{x}_i)$ is the feature mapping function that maps $\boldsymbol{x}_i$ from its input space to a feature space, ξ_i is the slack variable for outlier $\boldsymbol{x}_i$ that allows it to lie on the other side of the decision boundary (hyperplane), and $\nu \in (0, 1]$ is the regularization parameter that is an upper bound on the fraction of outliers and a lower bound on the fraction of support vectors.

If $\boldsymbol{w}$ and ρ solve this problem, we have the following decision function:

$$f(\boldsymbol{x}) = \mathrm{sgn}(\boldsymbol{w}^T\phi(\boldsymbol{x}) - \rho) \tag{9.14}$$

That is, if $\boldsymbol{w}^T\phi(\boldsymbol{x}) - \rho \geq 0$, $\boldsymbol{x}$ is declared as a normal event, otherwise, it is declared as intrusive.

To solve the quadratic programming (9.8), we introduce multipliers $\alpha_i, \beta_i \geq 0$, and get the Lagrangian:

$$L(\boldsymbol{w}, \boldsymbol{\xi}, \rho, \boldsymbol{\alpha}, \boldsymbol{\beta}) = \frac{1}{2}\|\boldsymbol{w}\|^2 + \frac{1}{\nu\ell}\sum_{i=1}^{\ell}\xi_i - \sum_{i=1}^{\ell}\alpha_i(\boldsymbol{w}^T\phi(\boldsymbol{x}_i) - \rho + \xi_i) - \sum_{i=1}^{\ell}\beta_i\xi_i \tag{9.15}$$

The following Karush–Kuhn–Tucker (KKT) conditions are necessary and sufficient conditions for the quadratic programming problem (9.8):

$$\frac{\partial L}{\partial \boldsymbol{w}} = 0, \frac{\partial L}{\partial \rho} = 0, \frac{\partial L}{\partial \xi_i} = 0, \quad i = 1, 2, \ldots, \ell, \tag{9.16}$$

$$\alpha_i \geq 0, \quad i = 1, 2, \ldots, \ell,$$

$$\beta_i \geq 0, \quad i = 1, 2, \ldots, \ell$$

According to the KKT conditions, set the derivatives equal to zero with respect to the primal variables $\boldsymbol{w}$, ξ_i, ρ, yielding:

$$\boldsymbol{w} = \sum_{i=1}^{\ell}\alpha_i\phi(\boldsymbol{x}_i), \alpha_i = \frac{1}{\nu\ell} - \beta_i \Leftrightarrow 0 \leq \alpha_i \leq \frac{1}{\nu\ell}, \sum_{i=1}^{\ell}\alpha_i = 1. \tag{9.17}$$

In Eq. (9.17), the $\boldsymbol{x}_i$ corresponding to $\alpha_i = 0$ is an inner point that is irrelevant to solving $\boldsymbol{w}$; the $\boldsymbol{x}_i$ corresponding to $0 < \alpha_i < \frac{1}{\nu\ell}$ is called a support vector, which is crucial to solve $\boldsymbol{w}$; $\boldsymbol{x}_i$ corresponding to $\alpha_i = \frac{1}{\nu\ell}$ is an outlier.

Now, we can rewrite the decision function (9.14) using the kernel function:

$$f(\boldsymbol{x}) = \text{sgn}\left(\sum_{i=1}^{\ell}\alpha_i k(\boldsymbol{x}_i, \boldsymbol{x}) - \rho\right) \tag{9.18}$$

In Eq. (9.18), $k(\boldsymbol{x}_i, \boldsymbol{x})$ is the kernel function with the following equivalent form:

$$k(\boldsymbol{x}_i, \boldsymbol{x}) = \phi(\boldsymbol{x}_i)^T\phi(\boldsymbol{x}). \tag{9.19}$$

For the OCSVM used in our detection model, we use the linear kernel:

$$k(\boldsymbol{x}_i, \boldsymbol{x}) = \boldsymbol{x}_i^T\boldsymbol{x} \tag{9.20}$$

Substituting Eq. (9.17) into Eq. (9.15), we can obtain the dual problem:

$$\begin{array}{ll} \min_{\alpha} & \frac{1}{2}\alpha_i\alpha_j k(\boldsymbol{x}_i, \boldsymbol{x}_j) \\ \text{s.t.} & 0 \le \alpha_i \le \frac{1}{\nu\ell}, \text{ for all } 1 \le i \le \ell \\ & \sum_{i=1}^{\ell} \alpha_i = 1 \end{array} \tag{9.21}$$

The answer to the dual problem (9.21) is also the answer to the primal quadratic programming problem (9.8). Furthermore, solving the dual problem is much easier and more feasible. We use the SMO (sequential minimal optimization) algorithm to solve the dual problem.

We know, any support vector $\boldsymbol{x}_i$ satisfies the equality:

$$\boldsymbol{w}^T\phi(\boldsymbol{x}_i) - \rho = 0 \tag{9.22}$$

So we can use the support vector $\boldsymbol{x}_i$, i.e., the corresponding α_i that satisfies $0 < \alpha_i < \frac{1}{\nu\ell}$, to recover ρ by the following formula:

$$\rho = \boldsymbol{w}^T\phi(\boldsymbol{x}_i) = \sum_{i=1}^{\ell} \alpha_i k(\boldsymbol{x}_j, \boldsymbol{x}_i). \tag{9.23}$$

9.3.4 Support Vector Data Description (SVDD)

Assume the training normal data are denoted as $\boldsymbol{x}_1, \boldsymbol{x}_2, \ldots, \boldsymbol{x}_\ell$, where $\boldsymbol{x}_i \in R^n, i = 1, 2, \ldots, \ell$. In SVDD, the normal class is mapped from the input space into a feature space via a mapping function $\phi(\boldsymbol{x})$. In this feature space, the normal class is denoted as

$$\phi(\boldsymbol{x}_1), \phi(\boldsymbol{x}_2), \ldots, \phi(\boldsymbol{x}_\ell), \tag{9.24}$$

where $\phi(\boldsymbol{x}_i)$ is the image of sample $\boldsymbol{x}_i$. The purpose of mapping function $\phi(\boldsymbol{x})$ is to render the patterns much more compact in the feature space than in the input space so as to improve the performance. Further, the inner products of two vectors in the feature space can be computed via a kernel function.

$$k(\boldsymbol{x}, \boldsymbol{x}_i) = \phi(\boldsymbol{x})^T\phi(\boldsymbol{x}_i), \tag{9.25}$$

where k satisfies the Mercer's theorem, $\phi(\boldsymbol{x})$ and $\phi(\boldsymbol{x}_i)$ denote two vectors in the feature space. In the feature space, support vector data description is used to determine the smallest sphere of radius $R > 0$ that encloses all the normal class approximatively as follows:

$$
\begin{aligned}
\min_R \quad & R^2 + \gamma \textstyle\sum_{i=1}^{\ell} \xi_i \\
\text{s.t.} \quad & \|\phi(\boldsymbol{x}_i) - \boldsymbol{o}\|^2 \le R^2 + \xi_i, \ \text{for all } 1 \le i \le \ell, \\
& \xi_i \ge 0, \qquad\qquad \text{for all } 1 \le i \le \ell
\end{aligned} \tag{9.26}
$$

where $\|\cdot\|$ means the Euclidean norm and $\boldsymbol{o}$ denotes the center of the sphere, ξ_i are slack variables to relax the constraints, γ is a parameter that specifies the trade-off between the sphere volume and the errors. $\sum_{i=1}^{\ell} \xi_i$ means the penalty term accounting for the presence of outliers. By introducing Lagrangian function [14], problem (9.26) is changed into

$$
\begin{aligned}
\min_{\boldsymbol{\alpha}} \ & \textstyle\sum_{i=1}^{\ell} \alpha_i k(\boldsymbol{x}_i, \boldsymbol{x}_i) - \sum_{i=1}^{\ell} \sum_{j=1}^{\ell} \alpha_i \alpha_j k(\boldsymbol{x}_i, \boldsymbol{x}_j) \\
\text{s.t.} \ & 0 \le \alpha_i \le \gamma, \qquad \text{for all } 1 \le i \le \ell \\
& \textstyle\sum_{i=1}^{\ell} \alpha_i = 1,
\end{aligned} \tag{9.27}
$$

where α_i for $i = 1, 2, \ldots, \ell$ is the Lagrange multipliers and problem (9.26) is a standard quadratic optimization problem. On the other hand, the samples $\boldsymbol{x}_i$ for which $\alpha_i \neq 0$ are called support vectors (SVs). Assume $\boldsymbol{x}_k$ is one of the SVs, and $0 < \alpha_k < \gamma$ holds true, R can be calculated as follows:

$$
\|\boldsymbol{x}_k - \boldsymbol{o}\|^2 = k(\boldsymbol{x}_k, \boldsymbol{x}_k) - 2\sum_{i=1}^{\ell} \alpha_i k(\boldsymbol{x}_k, \boldsymbol{x}_i) + \sum_{i=1}^{\ell}\sum_{j=1}^{\ell} \alpha_i \alpha_j k(\boldsymbol{x}_i, \boldsymbol{x}_k) = R^2 \tag{9.28}
$$

For a test pattern $\boldsymbol{x}$, it is assigned into the normal class if the distance between it and the sphere center is smaller than or equal to the radius R; on the contrary, pattern $\boldsymbol{x}$ is then classified as outliers as illustrated in Fig. 9.4.

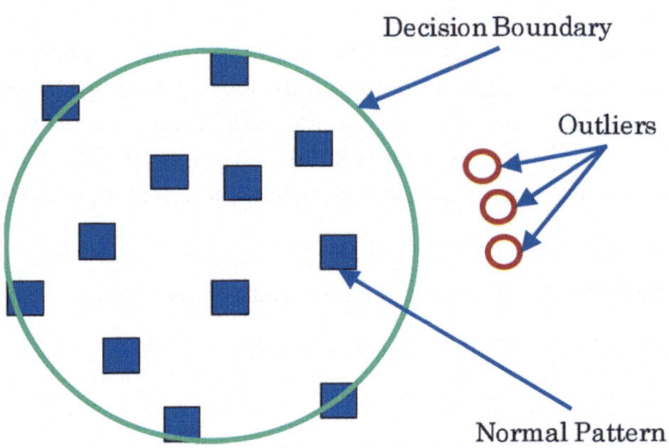

Fig. 9.4 Support vector data description for outlier detection

9.4 Experiments for Control System Communication Without Control Sequence Patterns

9.4.1 Experimental Setup

Our experiments used two kinds of control system communication data. One is water storage tank control system communication data which are supplied by Wei et al. [13]. Another one is gas pipeline control system communication data which are supplied Wei et al. [13]. Now, this chapter introduces these two kinds of control system communication data.

9.4.1.1 Water Storage Tank Control System Communication Data

Table 9.1 shows the features of water storage tank control system. The features are provided by Wei et al. [13]. We introduce the features briefly. The first feature and the second feature are the address numbers of the controlled equipments. These features show always same values in the normal condition. The features from the 3rd to the 6th are the memory information of the controlled equipment. The features from the 7th to the 11th are the commands to the controlled equipment. The 12th feature and the 13th feature are packet length of each command. The values of the 12th feature and the 13th feature do not change in the normal condition. The 14th feature shows the upper limit of water level. When the water level passes over the upper limit, alarm goes off. The 15th and the 16th features show the range of the normal water level. The 17th feature shows the lower limit of water level. When the water level underruns the lower limit, alarm goes off. The 18th and the 19th features show manual mode or automatic mode. When the value of the feature is 0, the control is off mode. When the value of the feature is 1, the control is manual mode. When the value of the feature is 2, the control is automatic mode. The 20th feature shows the running condition of the pump. When the value of the feature is 0, the pump is under suspension. When the value of the feature is 1, the pump is in operation. The 21st feature shows error rate of data transfer. The value of the feature is 0 in the normal condition. When the control system has DoS (denial of service)

Table 9.1 Features of water storage tank control system

1. Command address	2. Response address	3. Command memory
4. Response memory	5. Command memory count	6. Response memory count
7. Command read function	8. Command write function	9. Response read function
10. Response write function	11. Sub-function	12. Command length
13. Response length	14. HH	15. H
16. L	17. LL	18. Control mode
19. Control scheme	20. Pump	21. CRC rate
22. Measurement	23. Time	24. Result (label)

attack, the value of the feature changes. The 22nd feature shows water level in the tank, and the 23rd feature shows communication time interval. The values of these features are always changing. The 24th feature shows whether the data is attack or not. We describe details of the attack data in 9.4.1.3.

9.4.1.2 Gas Pipeline Control System Communication Data

Table 9.2 shows the features of water storage tank control system. The features are provided by Wei et al. [13]. We introduce the features briefly. In gas pipeline control system, the features from the 1st to the 13th are as same as the features of water storage tank control system. And also the 20th, 21st, 24th, 26th, and 27th features are as same as the features of water storage tank control system. The features from 14th to 18th show parameters of PID control. The 19th feature shows the set point of the pressure value in the pipeline. In normal condition, the feature always shows the same value. The 22nd and the 23rd features show the operating state of the valve. When the value of the feature is 0, the valve is under suspension. When the value of the feature is 1, the valve is in operation. The 25th shows the value of the pressure in the pipeline. The value is always changing.

If we only use the values of the features at a time t, this case is called "without sequence patterns."

9.4.1.3 Cyberattacks in Experiments

Wei et al generate the network traffic of a water storage tank control system and a gas pipeline control system. Wei et al suppose different kinds of attacks in the network traffic. In this chapter, we suppose the following 7 kinds of attacks:

1. NMRI (naive malicious response injection)
2. CMRI (complex malicious response injection)
3. MSCI (malicious state command injection)
4. MPCI (malicious parameter command injection)

Table 9.2 Features of gas pipeline control system

1. Command address	2. Response address	3. Command memory
4. Response memory	5. Command memory count	6. Response memory count
7. Command read function	8. Command write function	9. Response read function
10. Response write function	11. Sub-function	12. Command length
13. Response length	14. Gain	15. Rest
16. Deadband	17. Cycletime	18. Rate
19. Setpoint	20. Control mode	21. Control scheme
22. Pump	23. Solenoid	24. CRC rate
25. Measurement	26. Time	27. Result (label)

5. MFCI (malicious function command injection)
6. DoS (denial of service)
7. Recon (Reconnaissance)

We describe each attack briefly. NMRI attacks against the measurement. NMRI sends a crack measurement value to the water storage tank control system and the oil pipeline control system. CMRI is as same as NMRI and attacks against the measurement. CMRI sends a recent measurement value to the water storage control system and the oil pipeline control system, instead of sending a present measurement value which is a correct value. MSCI attacks against the control mode. MSCI sends a crack state of the controlled equipment. MPCI attacks against set points of the water storage tank control system or the gas pipeline control system. In the water storage tank control system, the set points are HH, H, L, LL. MPCI changes the set points such as a upper limit and a lower limit of water level or oil pressure. MFCI attacks against the sub-function of the water storage tank control system or the gas pipeline control system. MFCI sends a crack command to the controlled equipment. DoS attacks against the network and gives too much load by sending too many data. Recon is a reconnoiter attack.

9.4.2 Control System Communication Data

Among the published data are data with abnormal water level and pressure value as normal data. Therefore, normal data showing abnormal values were excluded.

9.4.2.1 Difference Between Measured Values *d*

We added the following features as the difference:

$$x = \text{measurement}(t) - \text{measurement}(t-1) \tag{9.29}$$

$$d = \begin{cases} 0 & (x = 0) \\ 1 & (\text{otherwise}), \end{cases} \tag{9.30}$$

where t represents time, and $t = \{1, \ldots, T\}$.

9.4.2.2 Alarm *e*

The value of measurement is assumed to be always within the set value($[H, L]$for water storage tank system and $[0, setpoint]$for gas pipeline control system) for both water storage tank system and gas pipeline control system. We added the following features as alarm:

$$e(Water) = \begin{cases} 0 \text{ } (L > \text{measurementor} H < \text{measurement}) \\ 1 \text{ (otherwise)} \end{cases} \tag{9.31}$$

$$e(Gas) = \begin{cases} 0 \text{ } (0 > \text{measurementorsetpoint} < \text{measurement}) \\ 1 \text{ (otherwise)} \end{cases} \tag{9.32}$$

9.4.2.3 Data Normalization

The upper and lower limits are set for measurement of water storage tank control system and gas pipeline control system. We performed normalization as follows. When the value of normalized measurement is 1, it should be the average value of measurement. The normalized measurement value decreases from 1 as the value of measurement deviates from the average of measurement value.

$$\text{normalized measurement value} = 1 - \left| 1 - \frac{\text{measurement value}}{\text{ave(measurement vale)}} \right| \tag{9.33}$$

9.4.2.4 ID Frequency

The data we used contains features that have no meaning to the value itself, such as ID. We calculated the frequency of the appearing ID and used the value of the frequency in the experiment. We calculate the frequency of the features from 1st to 21st of the water storage tank control system and the features of the 1st to 24th of the gas pipeline control system. We defined the ID frequency as follows:

$$\text{ID frequency} = \frac{\text{Number of data with the same value in the same feature}}{\text{Number of data to use}} \tag{9.34}$$

9.4.2.5 Training Data and Test Data

Next, we introduce training data and test data. In this experiment, we generate training data by extracting 10,000 data from normal data. And we generate test data by extracting 9500 data from normal data and 500 data from cyberattack data. We calculate ID frequency for the training data and the test data by using Eq. (9.34). Table 9.3 shows the number of used data in this experiment. The normal data satisfies $d = 0$ and $e = 0$.

Table 9.3 The number of used data

	Water	Gas
Normal	80,733	48,062
1. NMRI	9187	2763
2. CMRI	12,460	15,466
3. MSCI	1833	782
4. MPCI	3725	7637
5. MFCI	1320	573
6. Dos	1237	1837
7. Recon	34,002	6805

Table 9.4 Confusion matrix

	Normal	Attack
Normal	True negative	False positive
Attack	False negative	True positive

9.4.3 Evaluation Criteria

We briefly explain evaluation criteria to measure the performance. In this experiment, we adopt detection rate and false positive rate which are used by Wei et al. [13]. The test data are classified to one class in Table 9.4. The detection rate and false positive rate are presented by the following equations:

$$\text{Detection rate} = \frac{\text{True Positive}}{\text{True Positive} + \text{False Positive}} \tag{9.35}$$

$$\text{False positive rate} = \frac{\text{False Positive}}{\text{True Positive} + \text{False Positive}} \tag{9.36}$$

The detection rate and false positive rate are from 0 to 1. When the value of the detection rate 1, all attack data can be detected. When the value of the false positive rate is 0, detected attack data do not have normal data. In this experiment, we use the averaged detection rate and the averaged false positive rate for evaluation criteria.

9.4.4 Experimental Results

Table 9.5 shows the detection rate and the false positive rate of C4.5 and SVM for the water storage control system (Water) and the gas pipeline control system (Gas).

Table 9.5 Experimental results by binary classification methods

	C4.5				SVM			
	Water		Gas		Water		Gas	
	Detection rate False positive rate		Detection rate False positive rate		Detection rate False positive rate		Detection rate False positive rate	
1. NMRI	0.990	0.019	0.985	0.008	0.989	0.018	0.985	0.000
2. CMRI	0.988	0.007	0.971	0.000	0.988	0.007	0.972	0.000
3. MSCI	0.955	0.021	0.949	0.000	0.955	0.024	0.949	0.000
4. MPCI	0.985	0.009	0.987	0.004	0.984	0.009	0.987	0.004
5. MFCI	0.998	0.000	0.944	0.000	1.000	0.000	0.958	0.000
6. Dos	0.998	0.007	0.978	0.004	1.000	0.007	0.979	0.007
7. Recon	1.000	0.000	1.000	0.000	1.000	0.000	1.000	0.000

Wei et al applied C4.5 to detect cyberattacks. They adopted the network traffic data without our features which are proposed in Sect. 9.4.2. The experimental results in Table 9.5 adopted the network traffic data included our proposed features. The experimental results of C4.5 in Table 9.5 are better than Wei et al's experimental results. From these results, the features is very important for detecting cyber attacks in control systems. And we can find that the detection performance of C4.5 is as same as the detection performance of SVM from Table 9.5.

Table 9.6 shows the detection rate and the false positive rate of LOF and SVDD and OCSVM for the water storage control system (water) and the gas pipeline control system (gas). In this experiment, the number of training data, which are just normal data, is 10,000.

Table 9.7 shows the detection rate and the false positive rate of LOF and SVDD and OCSVM for the water storage control system (water) and the gas pipeline control system (gas). In this experiment, the number of training data, which are just normal data, is 10,000.

In this anomaly detection methods experiment, the parameter k of LOF is 500. The parameter ν of OCSVM is 0.001 and the parameter r of SVDD is 1.9. And OCSVM and SVDD adopt a linear kernel.

We compare Table 9.7 with Table 9.5. The detection performance of binary classification methods is better than the detection performance of LOF and SVDD. Especially, the detection performance of LOF is quite poor. It is hard to apply LOF to detect attacks in the control system communication. The detection performance of OCSVM is as same as the detection performance of C4.5 and SVM. OCSVM is promising to detect attacks in the control system communication because the control system communication network and the business communication network do not have a connection now and we have not have cyberattack data in the control system yet.

From Tables 9.6 and 9.7, we can observe the experimental results in Table 9.7 are better than the experimental results in Table 9.6. From this experimental result, we understand the more training data increase, the more the performance increase in anomaly detection methods.

Table 9.6 Experimental results by anomaly detection methods (No. of training data is 10,000)

	LOF				SVDD				OCSVM			
	Water		Gas		Water		Gas		Water		Gas	
	Detection rate	False positive rate	Detection rate	False positive rate	Detection rate	False positive rate	Detection rate	False positive rate	Detection rate	False positive rate	Detection rate	False positive rate
1. NMRI	0.999	0.876	0.988	0.868	1.000	0.535	0.984	0.640	0.999	0.000	0.984	0.006
2. CMRI	1.000	0.879	0.991	0.900	1.000	0.129	0.974	0.087	1.000	0.001	0.974	0.021
3. MSCI	0.998	0.903	1.000	0.805	0.994	0.018	0.947	0.028	0.995	0.110	0.950	0.014
4. MPCI	0.999	0.891	0.994	0.902	0.997	0.025	0.986	0.023	0.996	0.256	0.973	0.048
5. MFCI	1.000	0.941	1.000	0.937	1.000	0.008	1.000	0.008	1.000	0.001	0.994	0.046
6. Dos	1.000	0.920	0.995	0.915	0.984	0.311	0.959	0.009	1.000	0.000	0.983	0.055
7. Recon	1.000	0.903	1.000	0.872	1.000	0.697	1.000	0.622	1.000	0.252	1.000	0.280

Table 9.7 Experimental results by anomaly detection methods (No. of training data is 30,000)

	LOF				SVDD				OCSVM			
	Water		Gas		Water		Gas		Water		Gas	
	Detection rate	False positive rate	Detection rate	False positive rate	Detection rate	False positive rate	Detection rate	False positive rate	Detection rate	False positive rate	Detection rate	False positive rate
1. NMRI	0.117	0.000	0.995	0.942	0.999	0.155	0.986	0.443	0.999	0.000	0.987	0.007
2. CMRI	0.009	0.987	0.999	0.935	0.999	0.051	0.976	0.104	0.998	0.000	0.975	0.020
3. MSCI	0.320	0.804	1.000	0.926	0.994	0.004	0.949	0.010	0.993	0.000	0.948	0.008
4. MPCI	0.221	0.830	0.995	0.923	0.997	0.003	0.989	0.003	0.996	0.000	0.986	0.022
5. MFCI	0.034	0.983	1.000	0.934	1.000	0.005	1.000	0.001	1.000	0.000	1.000	0.023
6. Dos	0.222	0.408	0.997	0.925	0.996	0.137	0.941	0.002	1.000	0.000	0.987	0.022
7. Recon	0.000	1.000	1.000	0.941	1.000	0.545	1.000	0.402	1.000	0.000	1.000	0.073

9.5 Intrusion Detection Considering Sequences

This section describes the sequence characteristics of control systems' network traffic are very important for intrusion detection in control systems. And two probabilistic models, which can model the sequence characteristic of control systems' network traffic, are briefly explained in this section.

9.5.1 Control Sequences for Intrusion Detection

This session is the content described in the reference [8]. Intrusion detection systems used in general computer systems widely use signature based systems, which build a model based on available knowledge of attacks. This has proved to be effective in a general computer system environment. But for new and unknown cyberattacks, there is no data that can be used to build a model for those attacks. It is relatively easy to collect a large amount of normal communication data; therefore, anomaly detection using only the available normal data becomes effective. However, signature based systems are superior when detecting known attacks, so either both types of intrusion detection would be applied, or we assume that the basic attacks would be prevented with the other security measures.

Another feature characteristic to control systems is the control system sequence in the network traffic. In a typical control system, many of the control system devices would be controlled in a certain order. For example, a certain valve would have to be closed before another valve can be opened. This is in order to prevent any physical damage to the control system. Generally, a violation of such a sequence would be prevented at the operator console, for example, by not allowing the execution at the user interface, thus preventing any mistakes by the operator. There are usually also physical locks in the control system to prevent damage to the devices. However, if an attacker does not use the legitimate user interface and also eludes or disables the physical locks, there is a possibility that the attacker could cause damage to the control system. This would clearly violate the control system sequence, which should be visible in the network traffic, so if the violation could be detected by the intrusion detection system, it would help prevent damage to the system. Also, if the attacker does not have enough prior knowledge of the control system, the attacker may unknowingly violate the control system sequence. If the intrusion detection system is checking the sequences, such an attack would easily be detected.

With the features mentioned above, each sequence in the network traffic data is counted. Note that for the training, only the legitimate network traffic data of the system is used, making it adequate for control systems. After the training data is processed, the test data is processed with the same method, and if the data falls within a certain percentage of the minority data, the intrusion detection system alerts the data as an attack. The main contribution of our research is to propose the control sequences as a new feature for intrusion detection. And it is novelty for control systems to improve the performance of intrusion detection by using the control sequences.

To build models of control sequences, our research adopts HMM and CRF, which can satisfy the above constraints. Next, we briefly introduce the two probabilistic models we used for intrusion detection: HMM and CRF.

9.5.2 *Hidden Markov Model (HMM)*

The formal definition of a HMM is as follows:

$$\lambda = (A, B, \pi) \tag{9.37}$$

S is our state set, and Y is the observation set:

$$S = (s_1, s_2, \ldots, s_N), Y = (y_1, y_2, \ldots, y_N) \tag{9.38}$$

We define Q to be a fixed state sequence of length T, and corresponding observations O:

$$Q = q_1, q_2, \ldots, q_T, O = o_1, o_2, \ldots, o_T \tag{9.39}$$

A is a transition array, storing the probability of state j following state i. Note the state transition probabilities are independent of time:

$$A = \left[a_{ij}\right], a_{ij} = P(q_t = s_j | q_{t-1} = s_i) \tag{9.40}$$

B is the observation array, storing the probability of observation k being produced from the state j, independent of t:

$$B = [b_i(k)], b_i(k) = P(o_t = v_k | q_t = s_i) \tag{9.41}$$

π is the initial probability array:

$$\pi = [\pi_i], \pi = P(q_1 = s_i) \tag{9.42}$$

Two assumptions are made by the model. The first, called the Markov assumption, states that the current state is dependent only on the previous state, this represents the memory of the model:

$$P\left(q_t | q_1^{t-1}\right) = P(q_t | q_{t-1}) \tag{9.43}$$

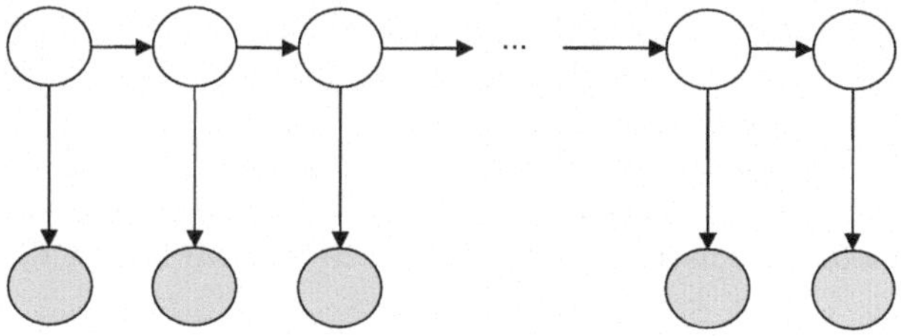

Fig. 9.5 Graphical structure of a chain-structured HMMs for sequences. The gray circles are observed variables and the circles are hidden variables

The independent assumption states that the output observation at time t is dependent only on the current state, it is independent of previous observations and states:

$$P\left(o_t|o_1^{t-1}, q_1^t\right) = P(o_t|q_t) \tag{9.44}$$

The problem of HMMs is to determine a method to adjust the model parameters

$$(A, B, \pi)$$

to maximize the probability of the observation sequence given the model. There is no known way to analytically solve for the model which maximize the probability of the observation sequence. In fact, given finite observation sequence as training data, there is no optimal way of estimating the model parameters [21]. We can, however, choose

$$\lambda = (A, B, \pi)$$

such as

$$P(O|\lambda)$$

is locally maximized using an iterative procedure such as Baum–Welch method (or equivalently the EM method) [22], or using gradient techniques. Figure 9.5 shows a directed graph of HMM.

9.5.3 Conditional Random Field (CRF)

A CRF may be viewed as an undirected graphical model, or Markov random field, globally conditioned on $\boldsymbol{X}$, the random variable representing observation sequences. Formally, we define $G = (V, E)$ to be an undirected graph such that there is a node $v \in V$ corresponding to each of the random variables representing an element Y_v of $\boldsymbol{Y}$. If each random variable Y_v obeys the Markov property with respect to G, then $(\boldsymbol{Y}, \boldsymbol{X})$ is a conditional random field. In theory the structure of graph G

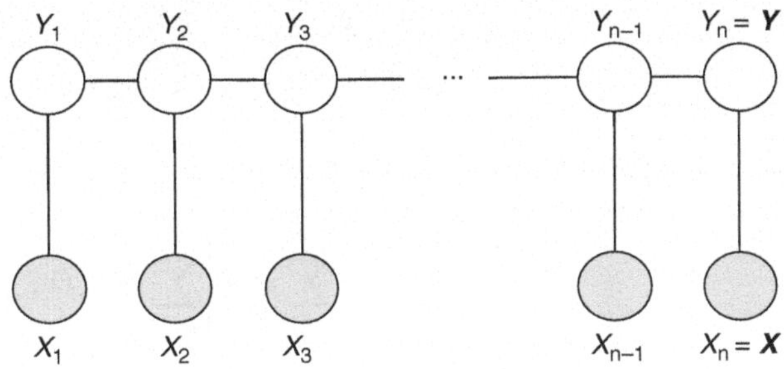

Fig. 9.6 Graphical structure of a chain-structured CRFs for sequences. The gray circles are observed variables and the circles are hidden variables

may be arbitrary, provided it represents the conditional independence in the label sequences being modeled. However, when modeling sequences, the simplest and most common graph structure encountered is that in which the nodes corresponding to elements of Y form a simple first-order chain, as illustrated in Fig. 9.6.

CRF is a probabilistic model which has a graphical structure. The input data is $\boldsymbol{x}$ and the estimated output data is $\boldsymbol{y}$. The general graphical models estimate a joint probabilistic distribution $P(\boldsymbol{x}, \boldsymbol{y})$. The general graphical models need to estimate a probabilistic distribution $P(\boldsymbol{x})$ to estimate the joint probabilistic distribution $P(\boldsymbol{x}, \boldsymbol{y})$. If the relation between input data is complicated, it is sometimes hard to estimate the joint probabilistic distribution. CRF estimates a conditional distribution $P(\boldsymbol{y}|\boldsymbol{x})$ directly.

CRF is computed by the following. The input data sequence set is X and the output label sequence set is Y. The input data sequence

$$x = \{x_1, \ldots, x_K\} \in X$$

and the output label sequence

$$\boldsymbol{y} = \{y_1, \ldots, y_K\} \in Y$$

consist of K elements. The conditional probabilistic distribution $P(\boldsymbol{y}|\boldsymbol{x})$ is defined by the next equation.

$$P(\boldsymbol{y}|\boldsymbol{x}) = \frac{1}{Z(\boldsymbol{x})} \exp\left\{\sum_{k=1}^{K} \lambda_k f_k(\boldsymbol{x}_t, \boldsymbol{y}_t)\right\} \tag{9.45}$$

where

$$\Lambda = \{\lambda_k, k = 1, \ldots, K\}$$

is a set of parameters. $f_k(\boldsymbol{x}_t, \boldsymbol{y}_t)$ denotes a characteristic function and is represented by the next equation.

$$\text{If } \quad \boldsymbol{y}_t = \boldsymbol{y}_k, \quad f_k(\boldsymbol{x}_t, \boldsymbol{y}_t) = \boldsymbol{x}_k$$
$$\text{and if } \quad \boldsymbol{y}_t \neq \boldsymbol{y}_k, \quad f_k(\boldsymbol{x}_t, \boldsymbol{y}_t) = 0 \tag{9.46}$$

$Z(\boldsymbol{x})$ is a partition function and represented by the following equation:

$$Z(\boldsymbol{x}) = \sum_{\boldsymbol{y}} \exp\left\{\sum_{k=1}^{K} \lambda_k f_k(\boldsymbol{x}, \boldsymbol{y})\right\} \tag{9.47}$$

where

$$\boldsymbol{x}^{(i)} = \{x_1^{(i)}, \ldots, x_T^{(i)}\}, i = 1, \ldots, N$$

is a sequence of the input data and to

$$\boldsymbol{y}^{(i)} = \left\{y_1^{(i)}, \ldots, y_T^{(i)}\right\}, i = 1, \ldots, N$$

and N is the number of data to generate a probabilistic model. Finally, the parameter λ is estimated by the maximum likelihood estimation based on the following likelihood function [23]:

$$l(\lambda_i) = \sum_{i=1}^{N} \log P\left(\boldsymbol{y}^{(i)} | \boldsymbol{x}^{(i)}\right) \tag{9.48}$$

where

$$\boldsymbol{y}^{(i)}, \boldsymbol{x}^{(i)}$$

are a set of training data

$$D = \left(\boldsymbol{x}^{(i)}, \boldsymbol{y}^{(i)}\right)_{i=1}^{N}$$

to generate probabilistic models [24].

9.6 Experiments for Control System Communication with Control Sequence Patterns

This session is described based on the reference [8].

9.6.1 *Experimental Setup*

Our experiments used two kinds of communication data. One is benchmark data which are supplied by DARPA [25, 26]. Another one is communication data of a model control system. The data are generated by our model system. Now, this chapter introduces these two kinds of communication data.

9.6.1.1 Benchmark Data (DARPA Data)

In order to evaluate intrusion detection methods for an information system, our experiments adopted DARPA benchmark data [25, 26]. From DARPA benchmark data, we randomly pick out 10,000 packets of normal communication data for training probabilistic models of normal communication sequences. This training data consists of only normal communication data. To generate the test data, we randomly pick out 10,000 packets of communication data, which include both unauthorized communication data and normal communication data, for evaluating the performance of some intrusion detection methods from DARPA benchmark data. Our experiments generated ten training datasets and ten test datasets. The features used as input data are shown in Table 9.8.

From DARPA benchmark data, we can extract source IP address, destination IP address, source port number, destination port number, protocol, data length, sequence number, identification number of field device, type of control command or state information, interval between packets, interval between use of field device, interval between control command or state information retrieval, and frequency of use of field device at a time t. And we can also extract values of these features of input data at a time $t + 1$. When the length of sequences is two, we use the values

Table 9.8 Features of input data for DARPA data

Features
Source IP address (IP address, MAC address)
Destination IP address (IP address, MAC address)
Source port number
Destination port number
Protocol
Data length
Sequence number
Identification number of field device
Type of control command or state information
Interval between packets
Interval between use of field device
Interval between control command or state information retrieval
Frequency of use of field device

of the features at a time t and the values of the features at a time $t + 1$. And when the length of sequences is three, we use the values of the features at a time t, time $t + 1$, and $t + 2$. If we only use the values of the features at a time t, this case is called "Without sequences." After all, when we consider two length of sequences, one packet consists of the values of the features at a time t and the values of the features at a time $t + 1$.

9.6.1.2 Control System Communication Data by Simulator

The model control system communication network used in the experiment is shown in Fig. 9.7 and was developed in our laboratory. Japanese electric power companies have two communication networks, which are the control system network and business network, which is defined as a type of business social network whose reason for existing is business networking activity. And these communication networks are disconnected physically. Therefore, it is assumed here that no connections exist between our control system network and the corporate business network. All the equipment, applications, and security measures in figure belong solely to the control system network, and the placement and settings mentioned are irrelevant to the corporate business network. The control system master server located in the control center is the direct control and data acquisition interface to the control system equipment, or field devices. To control the field devices located in the substation, the master server communicates with the server in the substation, which in turn sends the actual control signals to the field device. In this model system, the control system field devices are emulated in the same terminal as the substation server. The state information of the field devices and any responses to the control commands are sent through the substation server to the control system master server, and then sent to any communicating operator consoles, which in this case are located within the control center.

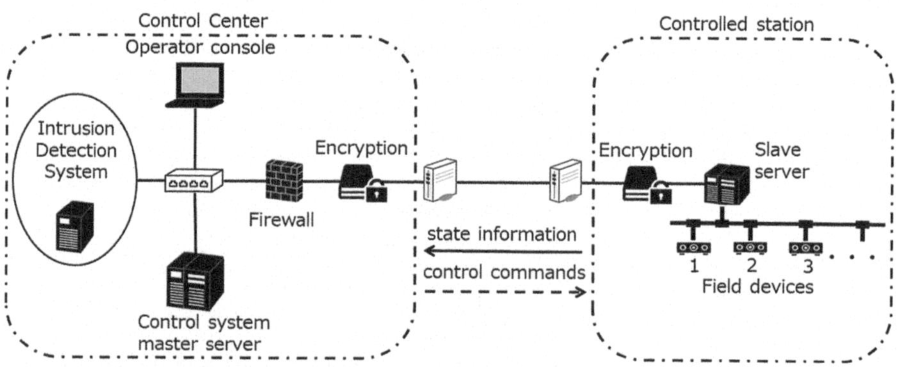

Fig. 9.7 An overview of the model system used for the experiment

Using the model control system, we collected 10,000 packets of normal communication data for training two probabilistic models. The features used from the data are shown in Table 9.9. The data was calibrated based on prior knowledge of the control system, to have values closer to 0 for less probable data.

Using the model control system, we collected 500 packets of unauthorized communication data for evaluating probabilistic models as intrusion detection systems. The unauthorized communication data are generated independently of the normal communication data by simulated cyberattacks, because we do not have any real cyberattacks for control systems. The model control system generated three types of the virtual cyberattacks. The three types of the virtual cyberattacks are the following:

1. Type 1: operational commands for non-existent controlled equipments
   This virtual cyberattack uses a controlled equipment ID which does not exist in the control system (see Fig. 9.8). In this case, an attacker does not have enough knowledge of the control system.
2. Type 2: iterative selection of different controlled equipments to virtually operate
   In this virtual cyberattack, a controlled equipment ID is a regularized ID. But

Table 9.9 Features used for training

Features	Description
Source IP address (IP address, MAC address)	Addresses belonging to devices in the system: 1
Destination IP address (IP address, MAC address)	Addresses belonging to the same network as the system: 0.5
	Addresses belonging to different network: 0
Source port number	Port numbers used in the system: 1
Destination port number	Port numbers reserved for use in the system: 0.5
	Port numbers not expected to be used in the system: 0
Protocol	Content specified to be used in the system: 1
Identification number of field device	Content not clearly specified to be used in the system: 0.5
Type of control command or state information	Content not clearly specified to be used in the system: 0.5
	Content clearly specified not to be used in the system: 0
Data length	Length within the range used in the system: 1
	Length out of the range used in the system: 0.5
Interval between packets	Time elapsed after previous packet
Interval between use of field device	Time elapsed after use of a certain field device
Interval between control command or state information retrieval	Time elapsed after use of the same control command or state information retrieval
Frequency of use of field device	Frequency of usage of a certain field device within the training dataset

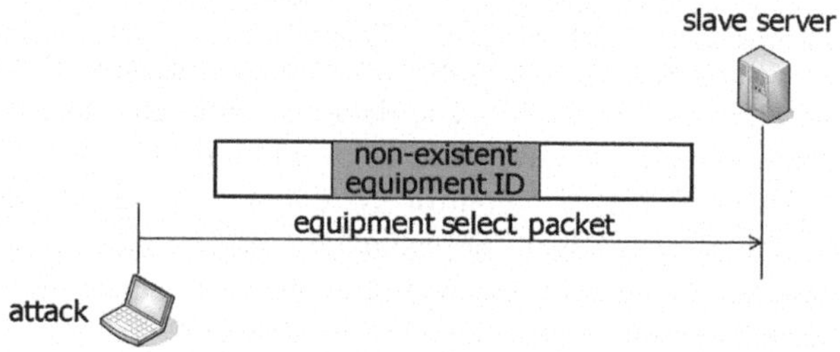

Fig. 9.8 Use of non-existent system equipment ID

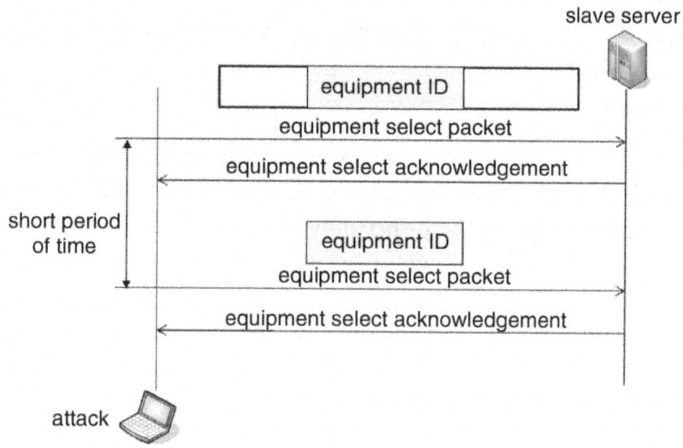

Fig. 9.9 Multiple and successive control select commands

the virtual cyberattack selects different controlled equipments iteratively (see Fig. 9.9). In this case, an attacker does not have enough knowledge of the control system and explore the control system architecture.

3. Type 3: iterative operational commands for a controlled equipment
   In this virtual cyberattack, operational commands are regularized commands. But this cyberattack sends regularized operational commands to a controlled equipment iteratively (see Fig. 9.10). In this case, the attacker wants to give a damage to controlled equipments.

9.6.2 Evaluation Criteria

In our experiments, we use the precision and the recall to evaluate the performance of the intrusion detection. The precision is represented by the next equation.

$$\text{Precision} = \frac{\text{True Positive}}{(\text{True Positive}) + (\text{False Positive})} \tag{9.49}$$

The recall is represented by the next equation.

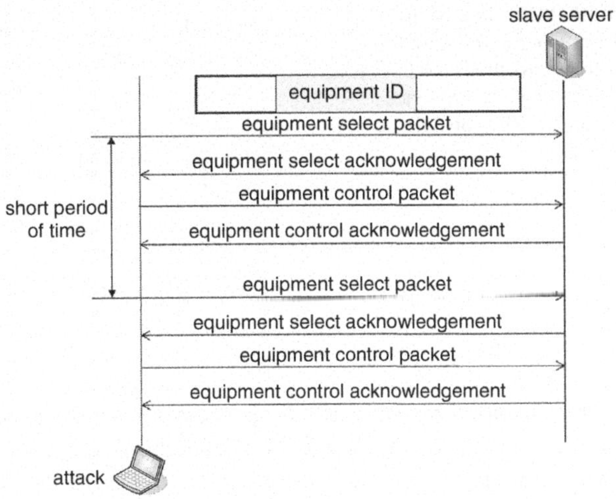

Fig. 9.10 Consecutive control commands during a short period of time

$$\text{Recall} = \frac{\text{True Positive}}{(\text{True Positive}) + (\text{False Negative})} \tag{9.50}$$

And we also use the F-measure to evaluate the performance and the F-measure is represented by the next equation.

$$F\text{-measure} = \frac{(1+\beta^2) \times (\text{Precision}) \times (\text{Recall})}{\beta^2 \times (\text{Precision} + \text{Recall})}, \tag{9.51}$$

where β denotes a relative importance between the precision and the recall and is usually set up 1 [27]. Therefore, we also set β up 1 in our experiments.

9.6.3 Experimental Results

This section presents two kinds of experimental results. One experimental results are based on DARPA benchmark data. Another experimental results are based on control system communication data.

9.6.3.1 Experimental Results for Benchmark Data (DARPA Data)

In our experiments, we apply training data, which are pick out 10,000 packets of normal communication data from DARPA data, and test data, which are pick out 10,000 packets of communication data, which consist of 500 unauthorized communication data and 9500 normal communication data, and two probabilistic models such as HMM and CRF. HMM and CRF train the sequence model of normal

Table 9.10 Precision, Recall and F-measure for different methods for DARPA data: Length of sequence denotes CRF probabilistic model

Method	Precision	Recall	F-measure
One-Class SVM	0.8933	0.8936	0.8934
SVDD	0.8976	0.8948	0.8962
Without sequences	0.8820	0.9465	0.9131
Rule based method	0.9034	0.8887	0.8960
HMM	0.8902	0.9180	0.9039
Length of sequence: 2	0.9021	0.9142	0.9207
Length of sequence: 3	0.9195	0.9219	*0.9216*
Length of sequence: 4	0.9189	0.9178	0.9183

2, 3 and 4 denotes the length of the considered sequence

communication by using 10,000 packets and set the existing probability of the most minority packet as the threshold of unauthorized packet. If the data falls within the threshold, the intrusion detection system alerts the data as an attack.

Table 9.10 shows the precision, recall, and F-measure of one-class SVM, SVDD (Support Vector Data Description), HMM and CRF and CRF without sequences information, and an empirical expertise of control systems (rule based method) [28]. The italic value in Table 9.10 shows the highest value of F-measure. One-Class SVM determines the data as an attack by using the distance from the origin. The training data were normalized from 0 to 1. The values of the minority data exist at the neighborhood of 0. In the experiments, one-class SVM adopts a linear kernel and set the value of the parameter ν to 0.00001. SVDD determines the data as an attack by using the radius of the optimal hyper-sphere. In the experiments, SVDD adopt a linear kernel.

The intrusion detection method without sequences shows that the performance of recall is high but the performance of precision is not high. The CRF based intrusion detection method shows the best performance of precision, recall, and F-measure when the number of input sequences is three. One-class SVM, SVDD, rule based method, and HMM show the same performance for DARPA data. The performance of the CRF based intrusion detection method depends on the length of sequences for DARPA data. But the differences between the performances are very small. F-measure of CRF based intrusion detection method is better than F-measure of one-class SVM, SVDD, rule based method, and HMM for DARPA data. The differences of the F-measure is small. DARPA data do not have the relation of sequences essentially. This denotes that it is not easy to improve the performance of intrusion detection by using the sequence information. However, CRF based on the sequences can improve the performance of intrusion detection. The improvement depends on distributed denial of service (DDoS) attacks.The DDoS attacks are sequential attacks to make system inaccessible by flooding the server's network and end user systems with fake generated traffic. Therefore, the DDoS attacks slightly improve the performance of CRF based on the sequences.

9.6.3.2 Experimental Results for Simulated Control System Communication Data

In our experiments, we apply training data, which are collected 10,000 packets of normal communication data from the model control system, and test data, which are collected 500 packets of unauthorized communication data from the model control system, two probabilistic models such as HMM and CRF. Table 9.11 shows the precision, recall, and F-measure of one-class SVM, SVDD (Support Vector Data Description), HMM and CRF and CRF without sequences information, and an empirical expertise of control systems (rule based method) [28]. The italic value in Table 9.11 shows the highest value of F-measure. From Table 9.11, the one-class SVM and SVDD show the same performance as CRF without sequences. The methods based on the control sequences show higher performance than the CRF without sequences. The two probabilistic models such as CRF and HMM achieve the higher performance than rule based method which uses the empirical expertise. In our experiments, CRF shows the higher performance than the other methods. Especially, when the length of the considered sequence is three, CRF achieves the highest performance in this experiment.

F-measure of CRF without sequences is lower than the other methods. We check the intrusion data which are passed by CRF without sequences. The intrusion data include typical intrusion data into the control system. The typical intrusion data can be detected by checking the sequence information. The proposed intrusion detection method based on CRF could detect most of intrusions in the model control system. But our proposed method could not detect few intrusions. In these few intrusions, an attacker intruded into the control system and made regular controls. It is impossible to detect intrusions by monitoring only control communication sequences when the attacker intruded into the control system and do nothing.

In our experiments, the performance of intrusion detection depends on the length of sequences. The network intrusion detection system has high frequent control sequences, which are generated by the monitored control system. So, the performance of intrusion detection depends on the length of sequences.

Table 9.11 Precision, recall, and F-measure for different methods for control system communication data: Length of sequence denotes CRF probabilistic model

Method	Precision	Recall	F-measure
One-Class SVM	0.9066	0.9006	0.9036
SVDD	0.9099	0.8960	0.9029
Without sequences	0.9016	0.8869	0.8942
Rule based method	0.9451	0.8931	0.9184
HMM	0.9523	0.9089	0.9301
Length of sequence: 2	0.9713	0.9198	0.9448
Length of sequence: 3	0.9839	0.9266	*0.9544*
Length of sequence: 4	0.9727	0.9279	0.9498

2, 3, and 4 denote the length of the considered sequence

9.7 Conclusion

The importance of cybersecurity has increased with the networked and highly complex structure of computer systems, and the increased value of information. Traditionally, control systems did not use networked communication systems. So, the cybersecurity was not important for the control systems. The networked control systems such as an intelligent distribution network system and so on are appearing and the cybersecurity will become very important for the control systems in the near future. However, we have few actual cyberattacks against the control systems. The intrusion detection should be developed by using only normal control system communication.

This chapter described intrusion detections for the control system communication without sequence patterns and for the control system communication with sequence patterns. In the former intrusion detection, we compared supervised machine learning based intrusion detection methods with unsupervised machine learning based intrusion detection methods. The supervised learning are C4.5 and support vector machine. And the unsupervised machine learning are local outlier factor, one-class support vector machine, and support vector domain description. We applied these intrusion detection methods to the water storage tank control system communication data and the gas pipeline control system communication data, and compared the differences in the performance. From the experimental results, we found two important points in the former intrusion detection part.

1. When we have enough training data, the intrusion detection performance of an unsupervised learning method can achieve the same intrusion detection performance of supervised learning methods.
2. The feature representation is very important for the intrusion detection of cyberattacks in control systems.

It is possible for the intrusion detection system to detect attacks that deviate from the typical system actions by using an unsupervised learning method with the appropriate features representation. In addition, the unsupervised learning based method does not need prior knowledge of attacks, which is the case for new, unknown cyberattacks.

In the later intrusion detection, we proposed the control sequences as a new feature for intrusion detection and compared HMM and CRF that reflect the sequence characteristics of network traffic into the intrusion detection system with one-class SVM, SVDD, and the rule based method. The learning only utilizes normal data, assuming that there is no prior knowledge on attacks in the system. The results show that the CRF, which identifies the sequences inherent in the system, could achieve better performance than HMM, one-class SVM, SVDD, and rule based method for the intrusion detection in our experiments. Therefore, the control sequences are very important feature to detect intrusion for control systems.

With the proposed CRF based on the sequence characteristics of network traffic, it is possible for the intrusion detection system to detect attacks that deviate from the typical system actions. In addition, this method does not need prior knowledge of attacks, which is the case for new, unknown cyberattacks.

Especially, the CRF based on the sequence characteristics of network traffic could achieve better performance than any other methods in our experiments. But F-measure of CRF is 0.9544. So, the proposed CRF could not detect some intrusions. In these some intrusions, an attacker intruded into the control system and made regular controls. It is impossible to detect intrusions by monitoring only control communication sequences when the attacker makes regular controls. We need to use another additional information to detect these intrusion.

Finally, the chapter showed two important things in the later intrusion detection part.

1. The experimental results show the sequence characteristics of network traffic is very important to detect some intrusion attacks in control systems.
2. We can expect that an intrusion detection method based on the sequence characteristics of network traffic in control systems will become a real world application in the near future.

References

1. National Security Agency, *Defense in Depth: A Practical Strategy for Achieving Information Assurance in Today's Highly Networked Environments*. http://www.nsa.gov/ia/_files/support/defenseindepth.pdf
2. SANS Institute, *Intrusion Detection FAQ*. http://www.sans.org/resources/idfaq
3. S. Cheung, B. Dutertre, M. Fong, U. Lindqvist, K. Skinner, A. Valdes, Using model-based intrusion detection for SCADA networks, in *Proceeding of the SCADA Security Scientific Symposium 2007* (2007), pp. 1–12
4. B. Moran, R. Belisle, Modeling flow information and other control system behavior to detect anomalies, in *Proceeding of the SCADA Security Scientific Symposium 2008* (2008)
5. A. Osareh, B. Shadgar, Intrusion detection in computer networks based on machine learning algorithms. Int. J. Comput. Sci. Netw. Secur. **8**(11), 15–23 (2008)
6. V. Chandola, A. Banerjee, V. Kumar, *Outlier Detection: A Survey*. University of Minnesota Technical Report TR 07-017
7. M. Kiuchi, Y. Serizawa, Security technologies, usage and guidelines in SCADA system networks, in *ICCAS-SICE 2009* (2009)
8. T. Onoda, Probabilistic models-based intrusion detection using sequence characteristics in control system communication. Neural Comput. Appl. **27**(5), 1119–1127 (2016)
9. J.S. Vinila, K.J. Jaya, Encrusted CRF in intrusion detection system, in *Artificial Intelligence and Evolutionary Algorithms in Engineering Systems* (Springer, New Delhi, 2015), pp. 605–613
10. W. Lee, S. Stolfo, M. Mok, A data mining framework for building intrusion detection model, in *Proceedings of the IEEE Symposium on Security and Privacy* (IEEE, Piscataway, 1999), pp. 120–132
11. R. Jain, N. Abouzakhar, A comparative study of hidden Markov model and support vector machine in anomaly intrusion detection. J. Internet Technol. Secured Trans. **2**, 176–184 (2013)
12. C. Alocious, N. Abouzakhar, H. Xiao, B. Christianson, Intrusion detection system using Bayesian network modeling, in *13th European Conference on Information Warfare and Security ECCWS 2014* (2014), pp. 223–232

13. W. Gao, *Cyberthreats, Attacks and Intrusion Detection in Supervisory Control and Data Acquisition Networks* (Department of Electrical and Computer Engineering, Mississippi State University, Mississippi State, 2013). Ph.D. thesis
14. V.N. Vapnik, *The Nature of Statistical Learning Theory* (Springer, London, 1998)
15. C.M. Bishop, *Neural Networks for Pattern Recognition* (Clarendon Press, Oxford, 1995)
16. N. Murata, S. Yoshizawa, S. Amari, Network information criterion-determining the number of hidden units for an artificial neural network model. IEEE Trans. Neural Netw. **5**(6), 865–872 (1994)
17. T. Onoda, Neural network information criterion for the optimal number of hidden units, in *Proceeding of ICNN'95-International Conference on Neural Networks*, vol. 1 (1995), pp. 275–280
18. J. Orr, K.-R. Müller, *Neural Networks: Tricks of the Trade*. LNCS, vol. 1524 (Springer, Heidelberg, 1998)
19. B.E. Boser, I.M. Guyon, V.N. Vapnik, A training algorithm for optimal margin classifiers, in *Proceedings of the fifth annual workshop on Computational learning theory* (ACM, New York, 1992), pp. 144–152
20. B. Schölkopf, A. Smola, R. Williamson, P.L. Bartlett, *New Support Vector Algorithms* (Royal Holloway College, University of London, London, 1998). NeuroCOLT Technical Report, NC-TR-98-031
21. L.R. Rabiner. A tutorial on hidden Markov models and selected applications in speech recognition. Proc. IEEE **77**(2), 257–285 (1989)
22. L.E. Baum, T. Petrie, G. Soules, N. Weiss, A maximization technique occurring in the statistical analysis of probabilistic functions of Markov chains. Ann. Math. Stat. **41**(1), 164–171 (1970)
23. CRF++, *Yet Another CRF Toolkit*. http://crfpp.sourceforge.net/
24. J. Lafferty, A. McCallum, F. Pereira. Conditional random fields: probabilistic models for segmenting and labeling sequence data, in *International Conference on Machine Learning* (2001)
25. R.P. Lippmann, J.W. Haines, D.J. Fried, J. Korba, K. Das, The 1999 DARPA off-line intrusion detection evaluation. Comput. Netw. **34**, 579–595 (2000)
26. DARPA, *Intrusion Detection Evaluation Data-set*. http://www.ll.mit.edu/mission/communications/ist/CST/index.html
27. D. Zhang, C. Leckie, An evaluation technique for network intrusion detection systems, in *InfoScale '06 Proceeding of the 1st International Conference on Scalable Information Systems* (2006)
28. M. Kiuchi, E. Ohba, Y. Serizawa, Customizing control system intrusion detection at the application layer, in *Proceeding of the SCADA Security Scientific Symposium 2009* (Digital Bond Press, Sunrise, 2009)
29. T. Onoda, Mai Kiuchi: analysis of intrusion detection in control system communication based on outlier detection with one-class classifiers, in *Proceedings of the 19th International Conference, ICONIP 2012, Doha, Qatar, November 12–15, 2012*. Part V, Lecture Notes in Computer Science, vol. 7667 (2012), pp. 275–282

Index

T. Suzuki et al. (eds.), *Design and Analysis of Distributed Energy Management Systems*, Power Electronics and Power Systems,
https://doi.org/10.1007/978-3-030-33672-1

The manufacturer's authorised representative in the EU is Springer Nature Customer Service Centre GmbH, Europaplatz 3, 69115 Heidelberg, Germany. If you have any concerns regarding our products, please contact ProductSafety@springernature.com

Printed and bound by CPI Group (UK) Ltd, Croydon, CR0 4YY
22/09/2026
02221057-0002